Accuracy and Calibration Failures in Science and Engineering

Scrivener Publishing
100 Cummings Center, Suite 541J
Beverly, MA 01915-6106

Publishers at Scrivener
Martin Scrivener (martin@scrivenerpublishing.com)
Phillip Carmical (pcarmical@scrivenerpublishing.com)

Accuracy and Calibration Failures in Science and Engineering

Michael D. Holloway

Scrivener Publishing

WILEY

This edition first published 2026 by John Wiley & Sons, Inc., 111 River Street, Hoboken, NJ 07030, USA and Scrivener Publishing LLC, 100 Cummings Center, Suite 541J, Beverly, MA 01915, USA

For more information about Scrivener publications please visit www.scrivenerpublishing.com.

Wiley Global Headquarters
111 River Street, Hoboken, NJ 07030, USA

For details of our global editorial offices, customer services, and more information about Wiley products visit us at www.wiley.com.

The manufacturer's authorized representative according to the EU General Product Safety Regulation is Wiley-VCH GmbH, Boschstr. 12, 69469 Weinheim, Germany, e-mail: Product_Safety@wiley.com.

AI disclosure statement: The author confirms that no artificial intelligence (AI)-generated content or AI-assisted writing tools were used in the preparation of this chapter (including summaries and the abstract). All text, analysis, and conclusions are the result of the author's original work.

Library of Congress Cataloging-in-Publication Data

ISBN 9781394417025

Cover image: Generated with AI using Adobe Firefly
Cover design by Russell Richardson

Set in size of 11pt and Minion Pro by Manila Typesetting Company, Makati, Philippines

Printed in the USA

10 9 8 7 6 5 4 3 2 1

This book is dedicated to Maurice Cusick and Fran Christopher because they are almost as big of Star Trek fans as I am!

Contents

Preface

I am a 'Trekkie,' in other words, a fan of Gene Roddenberry's *Star Trek*. Growing up in the 70s, *Star Trek* provided a glimpse of a universe that was completely new, fascinating, and, in some instances, horrifying. I remember a few other science fiction shows of the late 60s and 70s, such as *The Twilight Zone* (1959–1964) by Rod Serling, which would sometimes explore these ideas along with many others that brought about high interest (as well as nightmares!), and *The Outer Limits*, which was darker and more horror-leaning than *Star Trek*, with standalone episodes involving aliens, scientists, and futuristic tech. Like *Star Trek*, this often included strong moral or political subtexts. Then, there was *Lost in Space* (1965–1968), and while they encountered aliens and strange planets, I found the acting staff very annoying; however, they did provide the basis of examples of failure that ring true to this day! The best was (and still is) the *Doctor Who* series, which is just as old as I am and is still in production.

These shows produced a foundation of wonderment for science and the potential for the future. Without these concepts, I am certain I would have pursued a career in music instead of science and engineering. There is one thing that rang true to many of these episodes: there was always an element of failure involved. While this added to the overall storyline, I would hope certain lessons were learned.

In *Star Trek*, a tricorder reading was trusted without hesitation. Sensors identified anomalies, tracked life signs, and reported environmental hazards. But if a tricorder drifted off calibration, the crew would have been walking blind into disaster. That same blind trust in our own instruments has played out across industries with real consequences.

I wrote this book because calibration drift is not a theoretical nuisance. It has cost companies money, delayed projects, and in some cases put lives at risk. A pressure transducer off by a fraction can mask the start of a blowout. A thermocouple creeping away from true can send a turbine into conditions it was never designed to handle. In reliability work, those numbers drive decisions. If the numbers are wrong, the decisions are wrong.

The chapters are structured like case logs. Each failure is examined for what shifted, what went wrong, and what was done afterward. Some cases are minor, others catastrophic. They share one feature: assumptions were made that calibration was permanent, or that drift would be obvious. It rarely is. Drift hides in plain sight until the consequences are too big to ignore.

I use examples from aerospace, medical imaging, oil and gas, environmental monitoring, and even algorithm design. These are not science fiction, though at times they read like it. When a medical device fails calibration in practice, it creates the kind of cascading storyline you would expect from a holodeck malfunction. Except the outcomes are real and measurable.

This book is not about theory. It is about working with instruments that are imperfect in environments that are unforgiving. Calibration drift will occur. The task is recognizing it, building safeguards against it, and refusing to let silence or complacency bury it. Measurement integrity is not a slogan; it is a discipline. Without it, every chart, every trend, and every root cause analysis loses its foundation.

The pages that follow are not merely an account of technical missteps, but a study of how science and engineering can falter when accuracy is assumed rather than assured. Calibration and measurement integrity sits at the foundation of every modern industry such as aviation, energy, healthcare, environmental monitoring, and beyond. Yet again, seemingly minor drifts in standards, overlooked recalibrations, or unchallenged assumptions have led to failures with enormous human and economic costs.

This book was written because I have seen too many instances where these failures were avoidable. Over my career, I encountered situations where a measurement labeled as precise was misaligned. The deeper I investigated, the clearer it became when calibration shifts, so too does trust in data, and with it the safety of systems, organizations, and communities.

The work is structured to be both analytical and practical. Each chapter presents real-world failures, their root causes, and the corrective actions that could have prevented them. From drifting thermocouples in jet engines to mis-calibrated sensors in offshore oil wells, from false positives in clinical testing to algorithmic bias in artificial intelligence, the examples illustrate a common thread: unnoticed drift and unspoken doubt often open the door to systemic failure.

But this is not a book about failure for failure's sake. It is about resilience, traceability, and the cultural courage to confront uncomfortable truths. Embedded within these chapters are tools, checklists, and framework resources designed to help engineers, scientists, managers, and policymakers strengthen their organizations against the quiet erosion of accuracy.

I do not expect this book to end calibration failures. What I do hope is that it encourages conversation, raises awareness, and provides tangible steps toward better safeguards. If we are willing to question our data, to revisit assumptions, and to communicate changes openly, we can reduce the likelihood of repeating the same mistakes.

As with many lessons in engineering, humility is required: to admit that drift is inevitable, that standards evolve, and that our systems are only as trustworthy as the rigor we apply to them. In that sense, this book is not just a technical guide but also an invitation to foster a culture where accuracy is valued as both a scientific and ethical responsibility.

Michael D. Holloway
President, 5th Order Industry
214-450-7864
MichaelDHolloway63@gmail.com
5thOrderIndustry.com

1

Introduction: Today is a Good Day to Fail

Abstract
This book dissects failures in data integrity across industries like aerospace and healthcare, rooted in assumptions, misinterpreted data, and cultural blind spots. It examines how unnoticed analytical drift can lead to misinformation and disasters. Real-world case studies reveal institutional oversights and the need for cultural shifts to uphold measurement integrity. It provides a toolkit for preventing systemic failures through safeguards and design principles. Each chapter offers key concepts, thorough analyses, and practical tools to improve precision and reliability. The exploration stems from past laboratory experiences in addressing standard drifts and data integrity issues.

Keywords: Star trek, calibration, data integrity, systemic failure, analytical drift, case studies, human systems

In an episode of Star Trek: Deep Space Nine, the crew travels back in time to the events of The Original Series episode "The Trouble with Tribbles." Upon boarding the original U.S.S. Enterprise, they encounter Klingons who look radically different from their 24th-century counterparts, with smooth foreheads instead of the now-familiar ridged ones. When pressed on the discrepancy, Worf, ever stoic and proud, replies:

"We do not discuss it with outsiders."

It was a brilliant in-universe nod to a glaring inconsistency, one that most viewers understood was not a plot hole but a production limitation. Still, the show chose not to ignore the inconsistency. It acknowledged it with grace and a wink and later, even provided an in-universe explanation.

In science and engineering, we are not always so bold. When data does not align, when results contradict expectations, or when calibration standards quietly shift without notice, the reaction is often similar: We do not discuss it with outsiders. Why? Because exposing these errors could shake public trust. Because admitting uncertainty is seen as a weakness.

Michael D. Holloway. Accuracy and Calibration Failures in Science and Engineering, (1–4)

But sweeping errors under the rug is not a solution; it is a recipe for systemic failure.

Modern industries such as aerospace, healthcare, nuclear energy, environmental science, and more depend on the integrity of their data and the systems that interpret it. Yet again and again, catastrophic failures have occurred not from a lack of information, but from flawed assumptions, misinterpreted data, system drift, and cultural blind spots.

This book is a forensic exploration of what happens when analysis gets it wrong.

It begins with a microscopic case, a seemingly minor shift in a particle size calibration standard, and then scales up, examining how unnoticed analytical drift can snowball into widespread misinformation, flawed decisions, and even disaster. Along the way, we'll dissect real-world case studies, analyze historical failures, and highlight institutional oversights. But this is not just about technical missteps. It is about the human systems around them, the culture of blind trust in data, the silencing of dissent, the absence of challenge loops, and the reluctance to revisit assumptions.

You will find here not only the failures but also a toolkit for preventing them: safeguards, design principles, and cultural shifts needed to uphold measurement integrity and model resilience. If we wish to trust the output of our systems, we must be willing to examine their inputs, especially when they seem off. If we hope to learn from mistakes, we must resist the urge to ignore them.

"Today is a good day to fail" so long as we learn, adapt, and recalibrate.

In the pages that follow, I have attempted to identify profound failures, the reasons behind them, and the corrective actions proposed in response. There is no legitimate reason why these failures should be repeated, yet I believe they will be. This is primarily because we often avoid discussing such uncomfortable matters. This work is laid out to be accessible and instructive, enabling readers to understand and learn from the issues presented. Each chapter begins with key bullet points highlighting the most important concepts, followed by a thorough examination of the topic, and concludes with a comprehensive reference list. The appendix is filled with practical tools that can be implemented across various industries.

We do not yet have the ability to go back in time and rectify situations like in *Star Trek: The Next Generation*. However, we do have the power to communicate and learn from our mistakes. We should strive to make this not just a practice but a habit, for it is the key to success.

The genesis of this book was an offshoot of another work I authored, *The Analytical Chemistry Laboratory Companion*. In that book, I included comprehensive chapters on instrumental and wet chemistry analysis, identifying common pitfalls in data integrity and offering remedies to address them. That work arose from extensive efforts I undertook to improve precision, accuracy, repeatability, and reproducibility across a global laboratory network. One particular type of analysis involved measuring particle count populations within a sample. Years ago, I began hearing rumors and rumblings about standard drift—how the standards used by labs had shifted, calling into question the reliability of the data. The consequence? A measurement labeled as "4 microns" was no longer truly 4 microns. I never forgot that and recently decided to explore how and why such drift occurs.

That initial inquiry led me down a path to investigate other examples of calibration gone awry. Along the way, I found myself physically nauseated. It was not until I stepped back and considered the potential benefit of this work that I realized its importance. This exploration is necessary and valuable. Maybe, just maybe, it will help foster the kinds of conversations we too often avoid. We should discuss such matters.

2

The Shift in a Calibration Standard

Abstract
Traceability ensures measurement accuracy, reliability, and compliance, linking data to recognized standards, crucial in aerospace, pharmaceuticals, and food safety. Calibration verifies instrument accuracy against known standards, and traceability ensures global measurement consistency. Calibration without traceability risks errors and non-compliance. The shift from NIST SRM 2806 to SRM 2806a highlighted calibration challenges, affecting particle size interpretation and cleanliness codes. Consequences included incorrect equipment assessments and warranty disputes. Lessons include communicating calibration changes, normalizing historical data, and training staff on metrological updates. Key standards include ISO 11171 and ISO 4406, integral to calibration and cleanliness code formats.

***Keywords*:** Traceability, calibration, NIST SRM 2806, particle size distribution, ISO cleanliness codes, analytical drift, measurement integrity

Traceability and Calibration

Traceability is essential in science and engineering because it underpins the reliability, reproducibility, and accountability of measurements, materials, and outcomes. At its core, traceability connects a measurement or process back to a recognized reference standard, often maintained by national or international metrology institutes. This link provides verification that the data or process is accurate, ensuring that engineers and scientists can trust the results and confidently make decisions based on them. In research, traceability allows experiments to be replicated under the same conditions, which is fundamental to the scientific method. In engineering, it ensures that systems or products built in different locations or times will perform consistently.

Michael D. Holloway. Accuracy and Calibration Failures in Science and Engineering, (5–12)

Traceability also plays a crucial role in regulatory and legal compliance. Many industries, such as aerospace, pharmaceuticals, and food safety, require measurements and materials to be traceable to meet stringent standards set by regulatory bodies like the FDA or ISO. In the event of a dispute or audit, documented traceability can provide evidence of due diligence and data integrity. Moreover, traceability is a pillar of quality assurance programs and calibration systems. Instruments and processes must be calibrated and maintained using standards that are traceable through an unbroken chain of comparisons to national standards, which helps minimize measurement drift and error over time.

In failure analysis and root cause investigations, traceability enables professionals to trace back through records of measurements, materials, and procedures to identify what went wrong, thereby supporting continual improvement and preventive actions. Finally, traceability ensures global compatibility and facilitates international trade. Products tested or validated using traceable methods can be accepted across borders, promoting consistency, reducing redundancy, and enabling collaboration in multinational engineering and scientific endeavors. In all these ways, traceability forms the backbone of credible, consistent, and high-integrity work in science and engineering. Calibration and traceability are fundamentally related to measurement accuracy, quality assurance, compliance, and scientific integrity.

Calibration

Calibration is the process of comparing a measurement instrument or system to a known standard to ensure its accuracy. It's essential in any field where precise and reliable measurements are required, such as manufacturing, pharmaceuticals, environmental monitoring, and laboratory testing.

Traceability

Traceability means that the measurement results can be linked through an unbroken chain of calibrations, each with stated uncertainties, to national or international standards (e.g., NIST in the U.S. or SI units globally). This ensures consistency and comparability of measurements worldwide.

How They are Related:

- Traceability gives calibration its credibility. Without traceability, a calibration cannot be trusted beyond the local context.

- Traceability ensures that calibration results are globally recognized. If you calibrate a pressure sensor in Houston and another in Berlin, traceability ensures they measure identically because both refer to the same standard.
- Calibration without traceability risks errors and non-compliance. In regulated industries (like FDA, ISO 9001, or aerospace), traceable calibration is required to pass audits and maintain certifications.
- Both underpin decision-making. Every quality control chart, lab test, or manufacturing limit relies on the assumption that instruments are calibrated and that those calibrations are traceable.

Calibration Failure

In the early 2000s, NIST Standard Reference Material (SRM) 2806 was used widely to calibrate particle counters. However, over time, SRM 2806 was found to have limitations, particularly in terms of stability, accuracy, and repeatability.

- SRM 2806 was found to be unstable, and there were concerns about its accuracy and repeatability.
- In response, NIST introduced a new version, SRM 2806a, in 2011.
- This newer version had a different particle size distribution, especially for the ISO code boundaries (e.g., 4 µm, 6 µm, and 14 µm).
- Most importantly, SRM 2806a used different calibration techniques and traceability, including changes to how the equivalent spherical diameter was measured (e.g., changes in how edge detection and scattering light were interpreted).

In the early 2000s, NIST Standard Reference Material (SRM) 2806 was widely adopted as a benchmark for calibrating particle counters used in fluid analysis and contamination monitoring. This standard provided a basis for ensuring consistency and comparability of results across different laboratories and instruments. However, over time, SRM 2806 was found to have limitations, particularly in terms of stability, accuracy, and repeatability. These issues raised concerns within the scientific and industrial communities, prompting a reassessment of calibration practices for particle sizing and counting.

In response to these concerns, NIST released an updated version of the standard in 2011, known as SRM 2806a. This revised material featured a significantly different particle size distribution, particularly affecting the ISO cleanliness code boundaries at key micron sizes such as 4 μm, 6 μm, and 14 μm. These shifts had implications for interpreting and comparing particle count data, especially when transitioning from SRM 2806 to SRM 2806a in operational environments.

Beyond changes in particle size distribution, SRM 2806a introduced new calibration methodologies and traceability protocols. One of the most important improvements involved the way particle sizes were measured and defined. The method for determining equivalent spherical diameter was refined, with enhancements in edge detection and light scattering interpretation playing a critical role. These changes led to more robust and reliable calibration standards, supporting greater confidence in particle counter data across a range of applications.

ISO 11171 is the calibration standard for liquid particle counters (LPCs), typically used in lubricants and hydraulic fluid systems. It provides a method to calibrate optical particle counters using test aerosols or fluids and traceable reference materials.

What Went Wrong

- The change in calibration shifted the apparent particle sizes; the same sample would now show different ISO cleanliness codes.
- Many laboratories and users were unaware that the particle count numbers were no longer comparable year-to-year.
- The change was not immediately communicated clearly across all stakeholders, especially users in industrial and field settings.
- As a result, misinterpretation of oil cleanliness data occurred, with some systems appearing cleaner or dirtier than they actually were.

The transition from SRM 2806 to SRM 2806a brought about significant, yet initially underappreciated, consequences in the field of particle counting and oil cleanliness analysis. One of the most critical issues was that the recalibration altered the apparent size of particles. This meant that the same fluid sample analyzed before and after the switch could yield different ISO cleanliness codes, leading to a perceived change in contamination levels despite no actual difference in the sample itself.

Unfortunately, many laboratories, maintenance teams, and end-users were unaware of this shift in calibration standards. As a result, they continued comparing new particle count data to historical trends without realizing the data were no longer directly comparable. This lack of awareness was compounded by insufficient and inconsistent communication from calibration authorities and equipment manufacturers. Industrial users and field technicians, in particular, were often left without the necessary context to interpret the changes accurately.

The outcome was a widespread misinterpretation of oil cleanliness data. Some systems appeared to be significantly cleaner or dirtier than they actually were, leading to inappropriate maintenance actions, skewed trend analyses, and unnecessary concern or misplaced confidence about fluid system conditions. This episode underscored the importance of transparent communication, user education, and traceability whenever fundamental changes are made to metrological standards or calibration practices.

Consequences

- Reliability professionals sometimes make incorrect decisions about equipment health.
- Lubrication and filtration systems were adjusted unnecessarily or not adjusted when needed.
- In some cases, warranty disputes occurred between equipment users and OEMs over fluid cleanliness compliance.
- This also created confusion during root cause failure analysis, where historical data was incorrectly benchmarked against newer test results.

The consequences of the calibration shift were far-reaching and deeply impacted reliability-centered maintenance programs. With particle size boundaries effectively redefined, reliability professionals unaware of the change often made incorrect assessments of equipment conditions based on skewed cleanliness codes. Decisions about whether to take corrective action, replace components, or adjust maintenance schedules were sometimes based on flawed data interpretations, undermining the very purpose of predictive and proactive maintenance strategies.

Lubrication and filtration systems, which rely heavily on accurate particle count data for optimal performance, were also affected. In many cases, systems were unnecessarily upgraded, adjusted, or flushed in response to what appeared to be worsening cleanliness levels. Conversely, some systems in need of intervention were left unchecked because the new calibration made them appear cleaner than they actually were under the old standard.

Further complicating matters, the calibration shift led to disputes between equipment users and original equipment manufacturers (OEMs), particularly where warranty compliance hinged on documented fluid cleanliness. Users who believed they were maintaining fluids within the specification were surprised when post-calibration data suggested otherwise, leading to disagreements over whether warranty conditions had truly been violated.

This discrepancy also introduced confusion during root cause failure analyses. Historical oil cleanliness data, now measured against a fundamentally different standard, no longer provided a reliable benchmark for assessing trends or tracing the origin of a failure. As a result, investigators could draw incorrect conclusions, missing the real issues or wrongly attributing the root cause to perceived cleanliness problems. This situation highlighted the critical need for clear calibration traceability, effective communication of changes, and a robust understanding of how measurement standards impact long-term equipment health monitoring.

Lessons Learned

- Traceability and awareness: Any shift in calibration standards must be thoroughly communicated.
- Historical data normalization: Laboratories must flag and normalize data pre- and post-standard changes.
- Training and documentation: Maintenance and lab staff need regular updates about metrological changes that impact asset decisions.

The challenges surrounding the transition from SRM 2806 to SRM 2806a offer important lessons for the broader field of reliability and fluid analysis. Foremost among them is the necessity of traceability and awareness. When calibration standards change, especially those as fundamental as particle sizing, stakeholders across the entire value chain must be clearly and proactively informed. This includes not only laboratory personnel but also field technicians, reliability engineers, OEMs, and asset managers. Transparent communication ensures that everyone understands the implications and can respond appropriately.

A second key lesson is the importance of historical data normalization. Laboratories and data analysts must take care to flag results that straddle a calibration change and, where possible, apply normalization or correction factors to maintain continuity in trend analysis. Without this step, year-over-year comparisons become misleading, and data-driven decisions lose their integrity. Establishing a robust process for identifying and accounting for standard shifts protects the reliability of long-term datasets.

Lastly, training and documentation play a critical role. Maintenance personnel and laboratory staff need regular briefings and updated reference materials to keep pace with changes in metrology and test methodology. Calibration changes should be treated with the same seriousness as equipment upgrades or procedural changes, with clear documentation and targeted training. By embedding calibration literacy into the culture of maintenance and reliability programs, organizations can avoid costly missteps and make more informed, confident decisions about asset health.

Key ISO Standards Involved

- ISO 11171:1999 – Original standard for calibration using MTD and SRM 2806.
- ISO 11171:2016 – Updated to incorporate newer SRMs and more accurate calibration techniques.
- ISO 4406:1999/ISO 4406:2017 – Defines the particle cleanliness code format (e.g., 20/18/15), impacted by changes in SRM sizing.

References

- International Organization for Standardization (ISO), *ISO/IEC 17025:2017 General Requirements for the Competence of Testing and Calibration Laboratories*, 2017.
- National Institute of Standards and Technology (NIST), *NIST Special Publication 250-65: Guidelines for Evaluating and Expressing the Uncertainty of NIST Measurement Results*, 2019.
- Flack, D., Traceable Measurements: Back to Basics. *Qual. Prog.*, 36, 5, 70–75, 2003.

3

Technical Root Causes Behind Calibration Failures

Abstract

The shift in particle count calibration standards resulted from inaccuracies and instability in SRM 2806, driven by variable particle shape and refractive index in Medium Test Dust (MTD). Outdated calibration techniques lacked SI traceability, causing imprecise size thresholds. Advances in measurement technology, like SEM, revealed OPC misclassifications. NIST's recharacterization led to SRM 2806a, with improved test dust stability and validated particle sizes. Poor communication hindered SRM 2806a's adoption, with labs using legacy curves, delaying detection of the shift. The change corrected technological and material inconsistencies, exposing flaws in historical data.

Keywords: ISO medium test dust (MTD), optical particle counters (OPCs), calibration techniques, measurement uncertainty, particle geometry, batch-to-batch consistency, SI-traceable standards

The underlying cause of the shift in particle count calibration standards was the inaccuracy and instability of the original SRM 2806 due to several interrelated factors:

Limitations of the Original ISO Medium Test Dust (MTD)

- MTD used in SRM 2806 had variable particle shape and refractive index, which affected light scattering in optical particle counters (OPCs).
- Particle sizing relies on optical response, which can vary based on particle geometry, not just size.
- Over time, batches of MTD differed, introducing lot-to-lot inconsistency.

Michael D. Holloway. Accuracy and Calibration Failures in Science and Engineering, (13–18)

The root cause behind the shift in particle count calibration standards can be traced to the inherent inaccuracy and instability of the original SRM 2806, which stemmed from several interrelated issues, particularly the limitations of the original ISO Medium Test Dust (MTD) used in its formulation. MTD, though widely accepted at the time, introduced significant variability into the calibration process.

One major problem was that the particles in MTD exhibited irregular shapes and inconsistent refractive indices, both of which critically influenced how they interacted with light in optical particle counters (OPCs). Since these counters rely on light scattering to infer particle size, any variation in shape or optical properties could lead to skewed sizing results. OPCs interpret particle size based not solely on physical dimensions, but on how the particle modifies the path and intensity of scattered light, a process highly sensitive to particle geometry and composition.

Moreover, the production of MTD lacked strict batch-to-batch consistency. As new lots of dust were manufactured over time, subtle differences in particle characteristics accumulated, leading to lot-to-lot variability in calibration results. This inconsistency made it difficult to achieve reliable repeatability in particle counting across instruments and timeframes, undermining confidence in the standard. These foundational issues with MTD ultimately necessitated a re-evaluation of calibration methodology, culminating in the development of SRM 2806a with improved traceability, particle uniformity, and measurement rigor.

Outdated Calibration Techniques

- Original calibrations were based on comparisons to reference microscope slides or gravimetric methods, lacking traceability to SI units.
- OPCs were calibrated using uncertain edge detection and resolution, leading to imprecise size thresholds (e.g., 4 μm could be anywhere between 3.5 and 5 μm).

Another critical factor driving the shift away from SRM 2806 was the use of outdated calibration techniques that lacked the precision and traceability required for modern reliability and contamination control programs. Early particle counter calibrations were often based on visual comparisons to reference microscope slides or gravimetric measurements, neither of which offered strong traceability to SI units. As a result, the measurements were not only less rigorous but also difficult to validate or replicate across different instruments and laboratories.

Furthermore, optical particle counters (OPCs) of the time were calibrated using edge detection algorithms and optical resolution that left considerable room for uncertainty. The process of identifying a particle's "edge" based on how it scattered light introduced variability, especially for particles of irregular shape or variable transparency. This meant that reported particle sizes could differ significantly from their actual physical dimensions. For instance, what was labeled as a 4 μm particle could, in reality, be anywhere between 3.5 μm and 5 μm, depending on the specific OPC, its settings, and the interpretation of its optical signal.

These imprecisions in size thresholding directly impacted ISO cleanliness code classifications and trend analysis, undermining the reliability of fluid cleanliness assessments. The cumulative effect of ambiguous calibration techniques and poor size discrimination further underscored the need for an overhaul, prompting the adoption of more sophisticated, SI-traceable methods as implemented in SRM 2806a.

Advances in Measurement Technology

- By the 2000s, scanning electron microscopy (SEM) and more sophisticated laser and image-based methods revealed that particle counters were misclassifying particle sizes.
- For example, what had been classified as a "6 μm" particle under SRM 2806 was actually closer to 7–8 μm, depending on instrument design.

A third driving force behind the shift in particle count calibration standards was the emergence of advances in measurement technology, which exposed critical flaws in the original calibration methods. By the early 2000s, technologies such as scanning electron microscopy (SEM) and high-resolution laser and image-based particle sizing systems had matured to a point where they could provide far more accurate and detailed assessments of particle size and morphology than traditional optical particle counters.

These advanced tools revealed that many OPCs were misclassifying particle sizes, often underestimating the actual dimensions of particles due to limitations in their light-scattering assumptions and calibration references. A particularly telling example involved the "6 μm" particle size classification under SRM 2806: when re-examined using SEM or refined laser diffraction techniques, these particles were often found to be closer to 7–8 μm in true physical diameter. The discrepancy varied depending on the instrument design and how it interpreted the particle's optical signature.

This mismatch between perceived and actual particle sizes meant that the ISO cleanliness codes, which are heavily reliant on precise size boundaries, were based on flawed size thresholds. As a result, the industry faced a growing credibility gap in particle count data, a gap that could only be closed by recalibrating standards with modern, traceable, and scientifically rigorous measurement techniques. SRM 2806a was introduced as a direct response to these revelations, incorporating more accurate particle sizing and improved traceability to better align calibration with physical reality.

Metrology Demands for Traceability

- NIST, under pressure to provide SI-traceable standards, initiated a recharacterization program.
- Resulted in SRM 2806a, with:
  - Improved stability of test dust (better suspension and dispersion).
  - Particle size distributions validated by SEM, flow imaging, and optical sizing.
  - Better documentation and statistical analysis.

Finally, the shift in particle count calibration standards was driven by growing metrology demands for SI traceability. As industries placed increasing emphasis on precision, repeatability, and global standardization, institutions like NIST faced pressure to ensure that all reference materials adhered to the International System of Units (SI). In response, NIST launched a comprehensive recharacterization program aimed at improving the scientific rigor and reliability of its calibration standards.

This initiative culminated in the development of SRM 2806a, a significantly improved standard reference material that addressed the core deficiencies of its predecessor. One of the key upgrades was the enhanced stability of the test dust, achieved through better suspension and dispersion properties, reducing the tendency of particles to agglomerate or settle unevenly during testing. This alone greatly improved consistency in calibration results across different users and environments.

Additionally, the particle size distributions in SRM 2806a were rigorously validated using a combination of advanced techniques, including scanning electron microscopy (SEM), flow imaging analysis, and optical sizing methods. These technologies provided cross-validation of particle dimensions and ensured more accurate classification at critical ISO code boundaries.

NIST also incorporated more thorough documentation and robust statistical analysis into the release of SRM 2806a, giving laboratories and equipment manufacturers clearer insight into the reference material's properties, measurement uncertainty, and proper usage. This step marked a significant advancement in metrological quality, aligning fluid cleanliness calibration with modern scientific standards and reinforcing confidence in particle count data used for reliability, maintenance, and warranty decisions.

Poor Communication and Industry Uptake

- Transition to SRM 2806a was not well-publicized in field or lab settings.
- No industry-wide normalization guidance was issued for comparing pre/post-calibration data.
- Many labs continued using legacy calibration curves without adjusting, which delayed the detection of the shift.

Despite the technical improvements introduced with SRM 2806a, the transition was hindered by poor communication and slow industry uptake, which significantly blunted the impact of the updated calibration standard. One of the most critical missteps was that the change was not well publicized, particularly in the settings where it mattered most: field operations and routine laboratory environments. Many frontline users remained unaware that a fundamental shift in calibration had occurred, leaving them unprepared to interpret the resulting changes in particle count data.

Compounding this issue, no industry-wide normalization guidance was provided to help laboratories and end-users reconcile pre- and post-transition data. Without clear protocols for adjusting or re-benchmarking historical results, many organizations continued to compare new particle counts directly against legacy data sets. This created false trends, misdiagnosed fluid conditions, and disrupted long-term contamination monitoring programs.

Numerous laboratories continued using outdated calibration curves, either due to inertia, lack of awareness, or the absence of updated software and procedural guidance. This delayed the widespread detection of the calibration shift and prolonged inconsistencies across test results. The net effect was a fragmented and uneven transition, with critical reliability and maintenance decisions being made on misaligned data, a situation that could have been largely avoided through better communication, coordinated industry guidance, and proactive training initiatives.

The cause was ultimately a confluence of technological limitations, material inconsistency, and evolving metrological standards, which, when corrected, caused a step-change in calibration results. The shift wasn't due to a single "failure," but rather a long-overdue correction that exposed how historical data had been flawed for years.

References

- Easterling, R.G. and Harris, C.M., *The Role of Measurement Science in Standards and Calibration*, NCSL International, 1996. ISBN: 978-0945370192.
- Drubin, D.G., The quality of research and the role of calibration in data integrity. *Mol. Biol. Cell*, 25, 19, 2949–2951, 2014.
- BIPM (International Bureau of Weights and Measures), *The International System of Units (SI)*, (9th ed.). BIPM, 2020. ISBN: 978-9292212507.

4

Calibration Drift in Action: Industry-Specific Failures

Abstract

Thermocouple drift in aerospace and power generation industries, notably impacting Type K and Type N sensors, led to undetected temperature deviations in jet engines and turbines during the 1980s-1990s. High-temperature oxidation and unstable metallurgy caused signal drift, undermining reliability. Corrective actions included mandatory recalibration schedules, adoption of AMS 2750 standards, and use of drift-resistant materials. Enhanced sensor design with protective sheaths and redundancy systems improved reliability. Lifecycle tracking and documentation of thermal exposure became essential, elevating calibration to a critical function for ensuring accurate temperature data in mission-critical applications. Several other case studies are explored.

Keywords: Thermocouple drift, pressure transducers, pH electrodes, CT scanner dosimetry, mass standards, gas chromatography (GC), optical emission spectrometry (OES)

There have been a few notable incidents across various industries where calibration standards changed or drifted without timely communication or detection, leading to significant consequences. These examples are often used in metrology and quality training to highlight the importance of traceability, communication, and oversight. Here are some documented cases:

Thermocouple Drift in Aerospace and Power Industries

Industry: Aerospace, Power Generation

- Timeframe: 1980s–1990s (still periodically an issue)

Michael D. Holloway. Accuracy and Calibration Failures in Science and Engineering, (19–78)

- Issue: Type K and type N thermocouples drifted due to high-temperature oxidation and unstable metallurgy. Calibration was assumed stable until failures occurred.
- Impact: Jet engine and turbine over-temp conditions were undetected or misreported.
- Root cause: Lack of awareness of age and use degradation in field-installed sensors.
- Resolution: Mandatory periodic recalibration and requalification schedules.

In the aerospace and power generation sectors, thermocouple drift emerged as a critical reliability issue between the 1980s and 1990s, with residual challenges persisting into the present day. This problem primarily affected type K and type N thermocouples, which were widely deployed for high-temperature monitoring in jet engines, turbines, and combustion systems. Initially, these sensors were assumed to maintain calibration over extended use periods; however, this assumption proved dangerously flawed in high-temperature, oxidizing environments.

The core issue was that thermocouples experienced drift due to high-temperature oxidation, metallurgical instability, and grain growth. Type K thermocouples, in particular, were prone to chromium depletion in the positive leg (chromel), while type N, designed to be more stable, still exhibited degradation under extreme thermal cycling and long-term exposure. Over time, this led to significant changes in the thermoelectric voltage output, causing measurement errors that went unnoticed in the absence of periodic recalibration.

The impact was serious: engines and turbines operating above safe temperature thresholds went undetected or misreported, increasing the risk of component damage, reduced service life, and, in some cases, catastrophic failure. Because these sensors were embedded in field systems, often in harsh environments, the degradation was gradual and often attributed to external factors rather than sensor drift itself.

The root cause was ultimately traced to a lack of awareness around sensor aging and thermal degradation mechanisms. Engineers and operators lacked visibility into how long thermocouples had been in service, and there were no standard protocols requiring recalibration after extended use or exposure to extreme conditions.

The resolution came in the form of stricter regulatory oversight and industry best practices, including the implementation of mandatory periodic recalibration, sensor requalification schedules, and stricter lifecycle tracking. These protocols ensured that sensors were not only validated before installation but also systematically evaluated throughout their service life. This shift marked a major improvement in thermal instrumentation reliability and highlighted the broader need for ongoing metrological awareness in mission-critical applications.

There have been several well-documented and consequential incidents across diverse industries where calibration standards drifted or changed without adequate detection or communication, leading to serious operational, safety, and financial repercussions. These cases have become cornerstone examples in metrology and quality assurance training programs, reinforcing the critical importance of traceability, proactive communication, and robust oversight in calibration management. Among the most illustrative of these is the issue of thermocouple drift, which profoundly impacted the aerospace and power generation sectors during the late twentieth century and continues to pose challenges today.

Between the 1980s and 1990s, type K and type N thermocouples, which were widely used for high-temperature monitoring in jet engines, gas turbines, and other combustion-based systems, began to exhibit significant calibration drift. These thermocouples, often deployed in harsh oxidizing environments, were assumed to maintain their calibration over long service lives. However, this assumption proved dangerously inaccurate. Over time, these sensors underwent chemical and structural changes, especially at elevated temperatures. In type K thermocouples, the chromel leg experienced chromium depletion, which altered the material's thermoelectric characteristics. Type N, though developed to provide improved stability, still succumbed to grain growth and drift under severe thermal cycling conditions. These changes in material composition and structure led to shifts in the voltage output, directly affecting temperature readings.

The consequences of this undetected drift were profound. Engines and turbines were routinely operating at temperatures above their designed thresholds without this being accurately reported by the sensor systems. In high-performance environments, even minor temperature deviations can lead to accelerated component fatigue, reduced service life, and, in extreme cases, catastrophic failure. Because thermocouples are often embedded deep within systems and function without regular human interface, the degradation occurred silently. Operators often attributed anomalies to external factors or assumed calibration remained valid simply because the sensor appeared intact. The reality was that these instruments, though still generating output, were no longer providing accurate data, undermining the very purpose of their integration.

Investigations into these failures revealed that the root cause was not necessarily a flaw in the thermocouple design but a systemic oversight in lifecycle management. There was a widespread lack of understanding among engineers and technicians regarding the degradation mechanisms that affect sensor accuracy over time. Without detailed records of installation dates, service hours, or environmental exposures, it was impossible to predict when or if recalibration was needed. Moreover, industry protocols at the time did not mandate routine recalibration or replacement of in-service thermocouples, allowing degraded sensors to remain in critical systems far past their effective lifespan.

The response to these failures was transformative. Regulatory bodies and industry leaders introduced stricter guidelines requiring the periodic recalibration and requalification of thermocouples, particularly in mission-critical systems. Manufacturers and plant operators were urged and often mandated to implement robust tracking systems that recorded sensor deployment dates, operating environments, and service durations. Calibration Management Systems (CMS) became essential tools for ensuring compliance, traceability, and timely recalibration. Additionally, the industry began to shift toward more stable sensor technologies and to design systems with easier access for maintenance and recalibration. These changes marked a significant evolution in metrological practice, underscoring that the mere presence of a sensor is not sufficient; it must be known, trusted, and validated throughout its entire lifecycle.

Ultimately, the thermocouple drift incidents serve as a powerful reminder that calibration is not a one-time event but a continuous process. Without proactive measures to detect drift and communicate changes, even the most advanced instrumentation can become a silent liability. These lessons resonate beyond aerospace and power generation, informing best practices in pharmaceuticals, healthcare, manufacturing, and any field where precision measurement is mission-critical.

Root Cause

Thermocouples, particularly type K and type N, have long been the workhorses of temperature monitoring in jet engines and gas turbines. However, throughout the 1980s and 1990s, and persisting sporadically into the present, both aerospace and power generation sectors experienced critical failures tied to thermocouple drift. These temperature sensors, often deployed in high-heat zones such as turbine exhaust paths, experienced progressive signal drift due to high-temperature oxidation, grain growth, and metallurgical instability. In many cases, calibration was assumed to be stable for the life of the sensor, especially in inaccessible or embedded locations.

As a result, over-temperature conditions went undetected or were grossly underreported, leading to hardware degradation, unexpected shutdowns, and, in several cases, in-flight engine damage or accelerated creep fatigue in power turbines.

The root cause of these failures stemmed from multiple interrelated factors. First, there was a lack of awareness about thermocouple degradation mechanisms, particularly the impact of oxide layer buildup, chromium migration, and microstructural instability in type K thermocouples operating above 1000°C. Field-installed sensors degraded in accuracy over time, yet no periodic recalibration or requalification was enforced in many installations. Second, there was a pervasive overreliance on initial calibration, with operators assuming that pre-installation accuracy would persist under extreme cyclic thermal loads. This was a flawed assumption, especially in high-vibration, high-temperature environments. Compounding the issue was the absence of life-cycle tracking or cumulative usage monitoring for installed sensors, which meant that drift patterns were often only detected retrospectively after anomalous readings or failures occurred. Finally, inadequate documentation and calibration standards allowed many sensor installations to persist without formal requalification intervals.

Root Cause Summary: Thermocouple Drift in Aerospace and Power Industries

Factor	Description
High-temperature drift	Type K and N thermocouples degraded from oxidation and metallurgical shifts
Lack of in-field recalibration	Sensors were not periodically verified *in situ* after extended use
Overreliance on Initial Calibration	Installers assumed stability despite harsh operating environments
Missing lifecycle tracking	No system in place to monitor sensor age, hours of use, or thermal exposure
Unenforced requalification protocols	Absence of mandatory inspection or revalidation schedules

Corrective Action

As failures accumulated, both industries implemented significant corrective actions to mitigate future risks. The most impactful reform was the introduction of mandatory recalibration and requalification schedules, particularly for critical engine or turbine zone thermocouples. Organizations such as the FAA, NRC, and EPRI published technical guidance requiring replacement or recalibration after a specified number of operating hours or thermal cycles. This was supported by new standards such as AMS 2750 (Pyrometry), which imposed stricter documentation and drift tolerance limits for thermocouples used in aerospace applications. Sensor manufacturers also began incorporating drift-resistant alloys and protective sheaths for high-heat applications, while system integrators moved toward dual-sensor redundancy and the inclusion of diagnostic flags for signal anomalies. Furthermore, some facilities integrated automated calibration verification routines during shutdowns or scheduled maintenance windows, allowing early detection of drift before it compromised safety or performance. These combined actions elevated the role of calibration from a background task to a critical, tracked lifecycle function, helping to restore confidence in temperature data used for real-time control and condition monitoring.

Corrective Action Summary: Thermocouple Drift Mitigation

Area	Corrective action
Recalibration protocols	Mandated interval-based sensor recalibration or replacement schedules
Pyrometry standards	Adoption of AMS 2750 and equivalent for high-reliability thermal monitoring systems
Material improvements	Use of drift-resistant type N thermocouples, sheathed designs, and oxidation barriers
Redundancy and diagnostics	Dual thermocouple systems with built-in validation flags
Lifecycle tracking	Documentation of installation date, use hours, and thermal cycling exposure

References

- NIST Special Publication 250-80, Thermocouple Materials and Calibration Standards.
- ASTM MNL12, Manual on Thermocouples.
- NASA TM-2004-212902, Thermocouple Degradation in Harsh Environments.

Pressure Transducer Zero Drift (BP Oil Spill Related)

Industry: Offshore Oil & Gas

- Timeframe: 2010
- Issue: The Deepwater Horizon rig had pressure transducers calibrated onshore, but offshore recalibration protocols were inconsistent. Drifted sensors led to misreading well integrity.
- Impact: Contributed to failure to detect blowout conditions.
- Safeguard afterward: Mandatory offshore recalibration logs and traceability for safety-critical instruments.

In the offshore oil and gas industry, a critical instrumentation issue came to light during the 2010 Deepwater Horizon disaster, specifically the problem of pressure transducer zero drift. These transducers, vital for monitoring well integrity and detecting early signs of dangerous pressure anomalies, had been calibrated onshore prior to deployment. However, once installed on the Deepwater Horizon rig, the protocols for offshore recalibration and validation were inconsistent or lacking altogether.

Over time, environmental conditions such as temperature fluctuations, pressure cycling, mechanical stress, and electronic degradation can cause zero drift in pressure sensors, where the baseline measurement shifts away from the true zero point. In the case of Deepwater Horizon, this drift led to misinterpretation of pressure readings, making it appear as though well pressures were within acceptable limits when, in fact, warning signs of an impending blowout were present but obscured by faulty data.

The impact was devastating. The inability to detect escalating pressure accurately contributed to the failure to recognize and respond to the blowout conditions in time, playing a role in the chain of events that led to the explosion, massive oil spill, and tragic loss of life.

The root cause was a lack of traceability and verification for pressure transducers in the field, particularly for safety-critical systems. Assumptions that calibration remained stable after deployment proved to be dangerously optimistic in the harsh and high-risk offshore environment.

In the wake of the disaster, the industry implemented stringent safeguards, including mandatory offshore recalibration logs, traceability documentation, and more rigorous validation protocols for pressure transducers and other safety-critical instruments. These measures ensure that sensors are not only calibrated during commissioning but are also periodically verified and recalibrated under actual operating conditions, greatly enhancing the reliability of real-time well integrity monitoring.

In the aftermath of the 2010 Deepwater Horizon disaster, one of the critical instrumentation failures that surfaced was the issue of pressure transducer zero drift. These devices play an essential role in offshore drilling operations by providing continuous monitoring of well pressure and fluid dynamics. Accurate pressure readings are especially vital in identifying early warning signs of well-controlled issues, including kicks and potential blowouts. Prior to deployment on the Deepwater Horizon rig, the pressure transducers had been calibrated onshore under controlled conditions. However, once these sensors were installed and subjected to the rigors of the offshore environment, inconsistencies and gaps in recalibration protocols became evident. Offshore calibration verification was not universally mandated or systematically enforced, leaving safety-critical sensors vulnerable to drift over time without detection.

Pressure transducers, like all electronic measurement instruments, are susceptible to gradual zero-point shifts due to environmental stressors such as temperature cycling, pressure fluctuations, mechanical vibration, and internal component aging. In the context of offshore drilling, where high-pressure conditions, salinity, and extreme operational demands are routine, this drift can become significant. Zero drift alters the baseline reading of a sensor, making it appear that pressure values remain within a normal range even when they do not. In the case of Deepwater Horizon, this failure mode meant that vital indicators of mounting well pressure went unnoticed. The instruments falsely indicated stability in the wellbore, masking the early signs of the blowout that would eventually lead to catastrophic consequences.

The implications of this sensor failure were far-reaching. As pressure readings were relied upon to assess the safety and integrity of the well, misinterpreted data due to drift gave operators a false sense of security. This delayed crucial decisions and interventions that could have mitigated the impending blowout. The inability to detect and respond to the true well conditions in a timely manner directly contributed to the explosion, which resulted in the loss of eleven lives and one of the most significant environmental disasters in oil and gas history. The reliance on miscalibrated sensors without proper traceability or validation underscored a critical vulnerability in offshore safety protocols.

Root cause investigations following the disaster revealed a systemic flaw: a lack of robust traceability mechanisms for sensor calibration and a dangerous overreliance on initial calibration performed onshore. Offshore conditions differ markedly from the laboratory or staging environments where initial calibrations are carried out. Without a structured protocol for recalibrating and verifying sensor performance *in situ*, drift went unnoticed. Compounding this issue was the absence of a centralized calibration management system that could log and track the recalibration status of all safety-critical instruments.

In response to this failure, the offshore oil and gas industry instituted a suite of reforms aimed at ensuring the integrity of pressure measurement systems. Chief among these was the establishment of mandatory offshore recalibration and verification protocols for all pressure transducers associated with well control and safety systems. Operators are now required to maintain comprehensive calibration logs, enforce traceability through unique sensor identification, and perform periodic recalibration or validation under actual field conditions. These steps are designed not only to catch zero drift before it becomes a critical risk but also to foster a culture of continuous vigilance in instrumentation management. The pressure transducer issue in the Deepwater Horizon incident has since become a core teaching case in metrology, safety engineering, and quality management training, illustrating the profound consequences of overlooking sensor lifecycle accountability in hazardous environments.

Root Cause

The root cause of the pressure transducer calibration failure aboard the Deepwater Horizon rig can be traced to a combination of procedural gaps, flawed assumptions, and unaccounted environmental factors. Foremost was the lack of *in situ* recalibration; pressure transducers were installed offshore without subsequent verification, allowing zero drift to accumulate undetected over time. Compounding this oversight was an overreliance on initial calibration performed onshore; operators assumed that factory or dockside calibration remained valid in the demanding offshore environment, despite significant differences in pressure, temperature, and mechanical stress. This assumption proved critically flawed. Additionally, the absence of a centralized calibration traceability system meant that there was no consistent method to track sensor installation dates, calibration status, or validation history for safety-critical devices. As operations continued, environmentally induced sensor drift caused by thermal cycling, pressure fluctuation, and mechanical fatigue further degraded sensor accuracy, but these effects were neither monitored nor modeled in maintenance routines. Finally, the underlying calibration protocols were inadequate, lacking requirements for periodic field verification of sensor accuracy under actual operating conditions. The combined effect of these factors was a systemic vulnerability in well integrity monitoring, directly contributing to the failure to recognize critical pressure anomalies during the final hours before the blowout.

Factor	Description
Lack of *in situ* recalibration	Pressure transducers were not recalibrated or verified offshore after deployment, allowing undetected zero drift to persist.
Overreliance on initial calibration	Calibration conducted onshore was assumed to remain valid in the harsh offshore environment, which proved to be a flawed assumption.
Absence of traceability systems	No centralized or enforceable system to track calibration status, sensor installation dates, or validation history for safety-critical instruments.
Environmental-induced sensor drift	Offshore operational conditions, temperature, pressure cycles, and mechanical stress caused progressive sensor degradation that was not accounted for.
Inadequate calibration protocols	Standard operating procedures did not require or enforce periodic field verification of pressure transducers under actual operating conditions.

Corrective Action

Following the Deepwater Horizon disaster in 2010, a critical area of reform centered on the calibration and reliability of pressure transducers used in subsea and surface well monitoring systems. Investigators from the Chemical Safety and Hazard Investigation Board (CSB, 2016) and the Bureau of Safety and Environmental Enforcement (BSEE, 2012) found that drifted pressure sensors calibrated onshore but not reverified offshore contributed to misinterpreting well integrity tests, ultimately delaying the recognition of dangerous pressure anomalies. These errors were exacerbated by an absence of real-time redundancy and inadequate alarm thresholds.

To address these failures, the industry implemented a series of corrective actions designed to restore trust in sensor reliability and data interpretation. First, offshore *in situ* calibration verification became mandatory for safety-critical instruments, such as annular pressure and casing sensors. It was no longer acceptable to rely solely on factory or onshore calibration; sensors had to be re-zeroed under operational conditions and logged regularly. Standards like API RP 75 (API, 2013) and API Standard 53 (API, 2012) were updated to reflect these requirements.

Second, the industry adopted stricter traceability protocols, requiring calibration records to be linked to recognized national standards (e.g., NIST) and independently verified. Third-party inspection became a common requirement for high-risk systems, particularly blowout preventers (BOPs), where confidence in pressure readings was paramount (BSEE, 2012). To mitigate the risk of undetected drift, systems were redesigned with redundant pressure sensors programmed to cross-validate readings. If disagreement between sensors exceeded preset thresholds, automated alarms or system lockdowns would be triggered, an approach now embedded in revised versions of API 53.

In parallel, the industry reemphasized the human factors dimension, training rig personnel to critically evaluate sensor data rather than rely unquestioningly on digital readouts. This was especially relevant for interpreting negative pressure tests, which were misread in the hours before the blowout due to drift-compensated data and poor alarm logic (CSB, 2016; Paté-Cornell & Parker, 2012). In the years following the spill, the Safety and Environmental Management Systems (SEMS) rule (30 CFR 250 Subpart S) was enforced more rigorously by BSEE, mandating that operators demonstrate compliance with calibration protocols, data traceability, and alarm threshold validation (BSEE, 2012; API RP 75, 2013).

Finally, investment accelerated in low-drift sensor technologies better suited for the high-pressure, high-temperature (HPHT) conditions of deepwater wells. Research and development programs promoted sapphire-based and MEMS-based transducers, with longer calibration cycles and lower zero-offset error rates (NACE, 2012). These innovations aimed not only to improve measurement accuracy but also to reduce operator dependence on frequent recalibration under hazardous conditions.

Corrective Actions for Pressure Sensor Drift (Post-Deepwater Horizon)

Area	Corrective action
Calibration protocols	Offshore *in situ* checks; regular drift audits
Documentation	Traceable calibration records; third-party verification
System redundancy	Multiple sensors with alert triggers for drift
Human factors	Training in drift recognition and negative test interpretation
Regulatory oversight	SEMS audits and enforcement by BSEE
Sensor design innovation	Deployment of more stable, low-drift transducer technologies

References

- API, *API Standard 53 – Blowout Prevention Equipment Systems for Drilling Wells*, 2012.
- API, *API RP 75 – Safety and Environmental Management Systems (SEMS)*, 2013.
- BSEE, *Final Investigation Report on the Deepwater Horizon Blowout Preventer (BOP)*, 2012, https://www.bsee.gov/sites/bsee.gov/files/reports/inspection/bsee-final-report.pdf.
- CSB, *Final Report: Explosion and Fire at the Macondo Well – Deepwater Horizon Rig*, 2016, Report No. 2010-10-I-OS, https://www.csb.gov/macondo-blowout-and-explosion/.
- Paté-Cornell, M.E. and Parker, D., Warnings Unheeded: Engineering Analysis of the BP Blowout. *Risk Anal.*, 32, 9, 1404–1415, 2012.
- NACE International, *Corrosion Control and Instrumentation in Offshore Systems*, 2012.

The pH Electrode Calibration (Environmental Monitoring)

Industry: Environmental Labs

- Timeframe: Ongoing issue into the early 2000s

- Issue: Poor calibration tracking of pH meters using outdated or contaminated buffer standards.
- Impact: Regulatory samples were misreported; some industrial wastewater discharges were incorrectly certified as compliant.
- Resolution: EPA and ISO 17025 updated guidance for pH buffer expiry, electrode conditioning, and calibration traceability.

In the environmental laboratory sector, particularly those monitoring water quality, wastewater discharge, and regulatory compliance, the calibration of pH electrodes has long presented a hidden but significant source of error. Well into the early 2000s, many labs struggled with inadequate calibration practices, primarily due to the use of outdated, contaminated, or improperly stored buffer solutions and a lack of rigor in tracking electrode performance and calibration history.

pH electrodes are inherently sensitive to age, usage frequency, temperature changes, and contamination, and they require routine conditioning and proper storage to maintain accuracy. However, in many environmental labs, buffers were reused beyond expiration dates, stored improperly (e.g., left open or exposed to air), or contaminated through poor handling. Additionally, calibration records were often informal or missing, and regular verification against secondary standards was not always enforced.

The impact was significant, particularly when laboratories were testing EPA-regulated samples. In some cases, industrial wastewater discharges were incorrectly reported as compliant with environmental regulations based on faulty pH measurements. This undermined regulatory enforcement, allowed potential environmental harm to go undetected, and in some instances, led to costly investigations and remediation efforts when data quality was later called into question.

The resolution came through strengthened oversight from both regulatory and accreditation bodies. The U.S. EPA and ISO/IEC 17025 introduced updated guidance and requirements, including:

- Strict adherence to pH buffer expiry dates and proper storage protocols.
- Clear procedures for electrode conditioning, cleaning, and replacement schedules.

- Documentation of calibration traceability and the use of fresh, certified buffer solutions.
- Routine performance checks and drift assessments to ensure ongoing electrode reliability.

These improvements have reinforced the credibility of pH measurements in environmental testing and emphasized the critical role of calibration discipline and traceability in regulatory science. They also serve as a reminder that even seemingly simple instruments, like a pH meter, can undermine entire monitoring programs if not managed with proper metrological rigor.

In the field of environmental laboratory testing, especially for facilities tasked with monitoring surface water, groundwater, and industrial wastewater discharge, the calibration of pH electrodes has emerged as a persistent and underappreciated risk to data integrity. Through the 1990s and well into the early 2000s, numerous laboratories operated without robust protocols for calibration traceability, buffer solution management, or electrode maintenance. While pH measurement might seem simple or routine, its precision is foundational for regulatory compliance, chemical process control, and environmental protection. Improper calibration and maintenance of pH meters resulted in widespread inaccuracies that, in certain cases, had direct regulatory and environmental consequences.

The problem stemmed from a variety of common yet critical oversights. Chief among these was the widespread use of expired or contaminated pH buffer standards. Buffer solutions, typically pH 4.01, 7.00, and 10.01, are highly sensitive to contamination and require proper sealing, storage, and disposal after limited use. Yet many laboratories reused open containers well past their expiration dates, exposing them to atmospheric CO_2 and microbial contamination. In addition, electrode performance was often taken for granted; conditioning routines were inconsistent, replacement schedules were deferred, and drift verification against known standards was irregular. Compounding the issue, calibration logs were frequently informal or missing altogether, limiting the ability to reconstruct historical data quality or defend analytical results during audits or investigations.

These gaps in metrological control resulted in significant consequences. Environmental labs are frequently responsible for testing samples submitted for regulatory certification, many of which inform EPA enforcement decisions and environmental compliance reports. In several documented cases, wastewater discharges from industrial facilities were incorrectly reported as within permissible pH limits based on flawed measurements. This misreporting had downstream effects, including the undetected release of potentially harmful effluents into ecosystems and the avoidance of regulatory penalties based on invalid data. The inability to trace calibration events or demonstrate the use of certified buffers during audits led to retractions, mandatory retesting, and, in some cases, civil or administrative penalties.

Recognizing the scope and persistence of the issue, both regulatory agencies and accreditation bodies moved to strengthen oversight and guidance. The U.S. Environmental Protection Agency revised its QA/QC protocols to specifically require expiration date adherence for pH buffers, proper documentation of calibration events, and the implementation of standard operating procedures (SOPs) for electrode maintenance. Likewise, ISO/IEC 17025, the international standard for testing and calibration laboratories, incorporated explicit requirements for buffer solution traceability, electrode lifecycle management, and calibration documentation. These measures collectively marked a turning point in the quality culture of environmental laboratories. Facilities were encouraged, and often required, to move from informal practices to formal metrological systems capable of withstanding regulatory scrutiny.

Today, the calibration of pH electrodes in environmental labs is far more rigorous than in decades past, with improved practices leading to better measurement reliability, environmental stewardship, and data defensibility. The lessons learned continue to reinforce a fundamental truth in measurement science: even basic instruments, when mismanaged, can compromise entire analytical frameworks. The case of pH electrode calibration drift thus remains a central teaching point in laboratory quality assurance, emphasizing the intertwined roles of traceability, discipline, and calibration rigor in producing trustworthy results.

Root Cause

Throughout the 1990s and into the early 2000s, laboratories across North America and Europe increasingly encountered a persistent yet underreported problem: pH electrode miscalibration due to outdated, contaminated, or improperly stored buffer standards and neglected electrode conditioning practices. While the pH meter itself was often presumed accurate, poor calibration hygiene led to substantial deviations resulting in misreported pH values in regulatory samples. In some documented cases, industrial wastewater discharges were incorrectly certified as compliant, despite actual violations of environmental permits (EPA, 2004). The credibility of lab results and the integrity of environmental reporting came into question.

The root cause of this issue was multifaceted. First, many laboratories failed to maintain strict calibration tracking, reusing buffer solutions past their expiry dates or failing to verify buffer pH with secondary standards. Buffers exposed to air or improperly sealed degraded over time due to CO_2 absorption, causing the assumed pH value to drift, often unnoticed. Second, electrode conditioning and maintenance were frequently neglected; electrodes were used dry or improperly stored, resulting in sluggish response times and junction drift. Compounding this was a false assumption of instrument reliability; labs often trusted the pH reading without verifying calibration slope or electrode health. Many facilities also lacked formal calibration documentation systems, and procedures were inconsistently applied between technicians or shifts. Furthermore, buffer traceability to NIST or SI standards was often assumed but not verified, especially when labs relied on bulk-prepared buffers or low-cost solutions without certificates of analysis.

Root Cause Summary: pH Calibration Failures in Environmental Monitoring

Factor	Description
Expired or contaminated Buffers	CO_2 absorption and microbial growth altered buffer composition over time
Poor calibration tracking	Inconsistent records, unverified buffer traceability, and procedural drift
Electrode neglect	Electrodes stored dry, not rehydrated, or not conditioned before calibration
Overreliance on instrument readout	Readings trusted without checking slope, response time, or junction drift
Lack of traceability	Buffer standards not verified against NIST-traceable or SI-certified references

Corrective Action

As discrepancies became more evident in interlaboratory comparisons and EPA audits, regulatory and standards bodies implemented a series of corrective actions. The U.S. EPA issued updated guidance for laboratories operating under Clean Water Act permits, requiring stricter verification of pH buffer expiry, documentation of electrode maintenance routines, and the use of fresh, certified buffer solutions. Concurrently, ISO/IEC 17025 was revised to include more specific clauses regarding calibration traceability, including requirements for certified reference materials and documented calibration intervals (ISO/IEC 17025:2005; updated 2017). Labs were expected to maintain electrode calibration slope logs, reject calibration curves falling outside of expected tolerances (typically 90% to 105%), and record electrode serial numbers, buffer lot numbers, and calibration time/date on each analysis batch record.

In addition, training programs for laboratory analysts began emphasizing the importance of electrode conditioning, including proper hydration storage and cleaning procedures using acid/base soaks or enzymatic cleaners. New generations of benchtop pH meters introduced calibration timers and expiration alerts, prompting analysts to recalibrate at defined intervals. Quality assurance units now commonly perform random buffer cross-checks and require periodic audits of calibration logs. These actions helped transition pH measurement from a background task to a critical quality-controlled parameter in environmental compliance reporting.

Corrective Action Summary: pH Calibration Reforms in Environmental Monitoring

Area	Corrective action
Buffer quality assurance	Use of NIST-traceable buffers with expiration logs and lot tracking
Calibration documentation	Mandatory logs for electrode slope, buffer ID, and calibration timing
Electrode maintenance	Routine conditioning, cleaning, and storage in appropriate hydration solutions
Instrument improvements	Built-in timers, slope verification, and drift detection in modern pH meters
Regulatory alignment	EPA and ISO 17025 revisions enforcing traceability and procedural discipline

References

- U.S. Environmental Protection Agency (EPA), *Method 150.1: Measurement of pH in Water and Wastewater*, 2004.
- ISO/IEC 17025:2017, *General Requirements for the Competence of Testing and Calibration Laboratories*.
- NIST, *Standard Reference Materials for pH Buffers: SRM 186 and SRM 187 Series*, 2006.
- American Public Health Association (APHA), *Standard Methods for the Examination of Water and Wastewater*, 23rd, 2017.

CT Scanner Dosimetry Drift (Medical Imaging)

Industry: Medical/Radiology

- Timeframe: 2008–2012
- Issue: CT machines were delivering higher radiation doses than shown, due to calibration drift of dosimetry sensors and software updates that changed dose models without notification.
- Impact: Patients exposed to unintended radiation levels.
- Resolution: The FDA required manufacturers to disclose calibration/software changes and provide consistent dose monitoring tools.

Between 2008 and 2012, the medical imaging industry, particularly in the field of computed tomography (CT), faced a serious issue involving dosimetry drift, where the actual radiation dose delivered to patients exceeded the values indicated by the machine. This problem arose from a combination of calibration drift in internal dosimetry sensors and software updates that altered dose estimation algorithms without adequately notifying end users or recalibrating the system accordingly.

CT scanners rely on embedded sensors and computational models to estimate and display the dose a patient receives during a scan. Over time, however, dosimetry sensors can drift due to component aging, temperature effects, and repeated usage, causing the displayed dose to no longer reflect the true radiation output. The situation was further complicated when software updates intended to improve imaging efficiency or speed quietly changed dose models, sometimes increasing the effective dose without changing the displayed values.

The impact was alarming: patients were exposed to higher-than-intended radiation levels, often without the knowledge of the imaging technicians or physicians. In some cases, repeated scans led to cumulative exposure significantly above recommended safety thresholds, raising long-term concerns about radiation-induced risks, especially in vulnerable populations like children and oncology patients.

The root cause lay in a lack of transparency and traceability both in hardware calibration and software versioning. Clinicians assumed the systems were self-monitoring and accurate, unaware that internal changes had quietly compromised dosimetry reliability.

In response, the U.S. Food and Drug Administration (FDA) enacted critical safeguards. Manufacturers were required to disclose any calibration or software changes that could affect dose calculation and to document such changes in system logs. Additionally, the FDA mandated that CT systems include consistent, user-accessible dose monitoring tools, allowing operators to track, verify, and trend radiation exposure over time.

This incident underscored the need for rigorous metrological control, transparency, and communication in medical device calibration and highlighted how even high-tech, safety-regulated industries can be vulnerable when calibration and software versioning are not closely managed.

Between 2008 and 2012, the medical imaging field, particularly within radiology departments using computed tomography (CT), confronted a critical problem concerning dosimetry drift. CT scanners, which are routinely used to capture detailed cross-sectional images of internal organs and tissues, depend on built-in dosimetry systems to estimate and control the radiation dose delivered to patients. These systems consist of embedded sensors and dose modeling software that work together to calculate the effective dose per scan. However, during this period, a growing number of incidents revealed that some machines were delivering higher radiation doses than displayed, primarily due to drift in the calibration of internal dosimetry sensors and undisclosed software updates that modified dose calculation algorithms without operator awareness.

The issue of sensor drift is not unique to CT systems; it is a well-understood phenomenon in any environment that depends on long-term electronic sensor performance. In the case of CT scanners, dosimetry sensors are subjected to thermal cycling, electromagnetic interference, aging of internal components, and mechanical wear, all of which can lead to deviations from their original calibrated state. When drift is not identified and corrected through periodic recalibration, the output values can become progressively inaccurate. Compounding this hardware issue were software patches and upgrades issued by some manufacturers that, while designed to improve throughput, resolution, or scan speed, also altered the dose estimation algorithms. These changes affected how the machine calculated the displayed dose but were not accompanied by updated documentation or recalibration protocols, leaving medical personnel unaware of any shift in actual exposure levels.

The consequences were significant. In many documented cases, patients were inadvertently subjected to radiation doses exceeding clinical expectations, sometimes by a factor of two or more. Because the system displayed misleading dose values, clinicians continued to operate under the assumption that they were within safe margins. Particularly concerning were situations involving pediatric imaging or oncology patients, where multiple scans over time could result in cumulative radiation exposure that surpassed safe thresholds, increasing the long-term risk of radiation-induced conditions. Investigations and retrospective analyses conducted by hospitals and oversight bodies revealed that these errors were not caused by technician negligence but rather by systemic failures in calibration oversight and communication from equipment manufacturers.

The root cause of this issue ultimately pointed to a lack of traceability and transparency in both the hardware and software subsystems of CT scanner platforms. Calibration records for dosimetry sensors were not always maintained or required at the frequency necessary to detect drift. Furthermore, there were no regulatory mandates requiring vendors to notify users when a software update might affect radiation output. Clinicians, assuming the systems were fully self-monitoring and self-correcting, were left without the means to verify whether displayed dose values were still valid. This situation exposed a gap in medical imaging metrology, namely, the over-reliance on internal system integrity without independent verification or mandatory recalibration intervals.

In response to these findings, the U.S. Food and Drug Administration (FDA) and international radiological safety bodies issued updated regulations requiring that manufacturers disclose any calibration-affecting software changes and provide more robust, user-friendly tools for real-time dose monitoring. Additional guidance emphasized the need for routine verification of dosimetry accuracy, documented calibration schedules, and better operator training on system behavior following updates. These changes have led to the development of more transparent calibration tracking systems, built-in dose audit logs, and standardized software version control practices, ensuring that imaging centers maintain not only image quality but also patient safety through rigorous calibration discipline.

Root Cause

Between 2008 and 2012, the radiology community experienced a critical realization: computed tomography (CT) scanners were exposing patients to higher radiation doses than indicated, often without clinicians' or patients' awareness. The root of the problem was not a single failure but the convergence of dosimetry calibration drift, non-transparent software updates, and underreported changes in dose calculation algorithms. In several widely reported incidents, including those investigated by the U.S. Food and Drug Administration (FDA), patients undergoing brain perfusion CT scans received radiation levels that were four to eight times higher than intended (FDA, 2010). The scanners' control consoles continued to display normal or acceptable dose estimates, masking the error and allowing procedures to continue without intervention. As a result, patients reported hair loss, skin erythema, and, in some cases, potential long-term risks due to cumulative exposure.

The root cause of this widespread miscalibration was twofold. First, *in situ* dosimetry sensors embedded within CT systems experienced drift over time, especially in older machines or those with high scan volumes. These sensors responsible for measuring delivered radiation dose (typically *via* CTDIvol or DLP metrics) were not always routinely recalibrated against national standards and discrepancies between reported and actual output accumulated. Second, during this same period, manufacturers issued software updates that modified image acquisition protocols or dose estimation algorithms without consistently alerting users to the implications of dosimetry. Changes to tube current modulation, reconstruction techniques, or patient size modeling altered actual delivered doses, but the graphical user interfaces (GUIs) continued to reflect outdated dose metrics. There was no uniform standard requiring disclosure of how updated algorithms would affect radiation output or displayed values. In many clinical settings, these shifts were only discovered retrospectively after abnormal patient reactions prompted the investigation.

Root Cause Summary: CT Scanner Dosimetry Drift

Factor	Description
Sensor calibration drift	*In situ* dosimeters degraded over time without periodic recalibration
Software update effects	Dose algorithms were updated without transparent documentation or clinician alerts
GUI/display inconsistency	User interfaces continued showing outdated or misleading dose values
Lack of audit or verification	No routine checks compared displayed vs. actual dose output
Procedural overconfidence	Technologists trusted dose readouts without independent verification

Corrective Action

The corrective actions that followed were significant and industry-wide. The FDA issued multiple safety communications starting in 2010, requiring CT scanner manufacturers to clearly disclose software and dose algorithm changes and to document their effect on dose estimation metrics (FDA, 2010). This led to a push for standardization of dose reporting, including widespread adoption of Radiation Dose Structured Reporting (RDSR) and integration of dose data into Picture Archiving and Communication Systems (PACS). Additionally, the American College of Radiology (ACR) and the Joint Commission mandated routine dosimetry audits, requiring that facilities periodically verify actual radiation output using external phantoms and calibrated dosimeters.

CT scanner vendors were also required to include dose monitoring dashboards, alerts for high cumulative exposure, and user-controlled limits for protocols. Modern scanners now provide real-time dose estimation, auto-adjusted for patient size and scan parameters, with historical tracking to flag excessive patterns. Further, in 2012 the Image Gently and Image Wisely campaigns emphasized radiation safety in pediatric and adult imaging, respectively promoting technician training, optimization of scan protocols, and documentation of dose per patient.

Perhaps most critically, these events reinforced that software is a calibration-dependent system component, not merely a passive controller. Any change in software that affects dose computation must be treated as a calibration-relevant event, demanding traceability, user awareness, and formal revalidation. These reforms have since re-established CT dosimetry as a fully auditable and safety-critical aspect of medical imaging.

Corrective Action Summary: CT Dosimetry Reforms

Area	Corrective action
Calibration verification	Mandatory recalibration of *in situ* dosimetry sensors with traceable external instruments
Software update transparency	FDA-mandated disclosures for algorithm changes impacting dose estimation
Real-time dose monitoring	Implementation of dashboards, alerts, and historical tracking in scanner UIs
Structured dose reporting	Adoption of Radiation Dose Structured Reports (RDSR) integrated into PACS systems
Education and Optimization	ACR/Joint Commission directives and campaigns (Image Gently/Wisely) promoting safe practice

References

- U.S. Food and Drug Administration (FDA), *Radiation Overexposure Reports in CT Brain Perfusion Scans*, 2010, https://www.fda.gov/radiation-emitting-products.
- American College of Radiology (ACR), *ACR Technical Standard for Diagnostic Medical Physics Performance Monitoring of CT Equipment*, 2011.
- Joint Commission, *Sentinel Event Alert #47: Radiation Risks of Diagnostic Imaging*, 2011.
- Mahesh, M., CT Dose and Optimization: Understanding the Concepts. *AJR Am. J. Roentgenol.*, 199, 3, 612–613, 2012.
- Brink, J.A. and Amis, E.S., Image Wisely: A Campaign to Increase Awareness of Adult Radiation Protection. *Radiology*, 257, 3, 601–602, 2010.

Mass Standards Drift (Pre-Kilogram Redefinition)

Industry: Fundamental Metrology/All industries relying on mass

- Timeframe: 1889–2018

- Issue: The International Prototype Kilogram (IPK), a platinum-iridium cylinder in Paris, lost mass over decades relative to national copies.
- Impact: All mass calibrations worldwide were potentially traceable to a drifting physical object.
- Resolution: The kilogram was redefined in 2019 using the Planck constant, eliminating reliance on a physical artifact.

From 1889 to 2018, the foundation of all mass measurements worldwide rested on a single physical object: the International Prototype Kilogram (IPK), a carefully crafted platinum-iridium cylinder stored under controlled conditions in Sèvres, France. This artifact served as the definitive reference for the kilogram, with all national mass standards and industrial calibrations traceable back to it. However, over more than a century, a profound issue emerged: the IPK itself was drifting.

Periodic comparisons between the IPK and its officially sanctioned copies, held by national metrology institutes around the world, revealed that the IPK was losing mass relative to its replicas by an estimated 50 micrograms over several decades. This drift, though minute in absolute terms, posed a major problem for high-precision industries and scientific research, where even microgram-level discrepancies could impact quality control, pharmaceutical dosing, materials research, and advanced manufacturing.

The root cause of the drift was likely due to surface contamination, cleaning procedures, and molecular changes that subtly altered the IPK's mass over time. Because all mass measurements were tied to this single artifact, its instability introduced systemic uncertainty across global calibration chains, an unsustainable risk in an era of ever-increasing precision requirements.

The resolution came in 2019 when the scientific community redefined the kilogram in terms of a fundamental constant of nature, the Planck constant (h). This was made possible by advances in quantum metrology and the development of the Kibble balance, which relates mechanical power to electromagnetic energy. By fixing the value of the Planck constant and linking mass to reproducible quantum phenomena, the new definition freed the kilogram from dependence on a physical object.

This historic shift not only eliminated the issue of artifact drift but also marked a profound evolution in metrology, transitioning from reliance on physical standards to a purely constant-based system. It ensured long-term stability, universality, and precision in mass measurements, benefiting every industry and laboratory that depends on accurate mass calibration.

From 1889 until its historic redefinition in 2019, the global system of mass measurement relied on a single, tangible artifact: the International Prototype Kilogram (IPK). Housed at the International Bureau of Weights and Measures (BIPM) in Sèvres, France, the IPK was a platinum-iridium cylinder that served as the definition of one kilogram. All other mass standards, from national metrology laboratories to factory floor calibration weights, were calibrated directly or indirectly against this one object. For over a century, the IPK represented the ideal of mass stability. However, over time, what was once considered an immutable reference came under scrutiny as evidence mounted that the IPK's mass was not as constant as once believed.

Periodic verifications of the IPK against its official copies, distributed to national metrology institutes around the world, began to reveal an unexpected and troubling trend: the IPK appeared to be losing mass relative to its replicas. While the discrepancy was small, about 50 micrograms over several decades, it was significant in the world of high-precision measurement. Industries such as aerospace, pharmaceuticals, and materials science often rely on nanogram-to-microgram resolution for research, production, and quality control. A drift of this magnitude, originating at the highest level of the calibration chain, propagated uncertainty throughout global systems, affecting everything from legal trade weights to sensitive laboratory balances.

The root causes of this drift were likely multifaceted. Surface contamination, micro-scale wear, cleaning-induced abrasion, and subtle molecular rearrangements could each have played a role in altering the IPK's mass. The problem was not that the IPK was uniquely unstable, but that it was a singular physical object upon which the entire concept of mass was based. Any change, no matter how small, introduced systemic uncertainty across every standard that relied on it. As science and industry demanded ever-greater levels of accuracy, reliance on a physical artifact became increasingly untenable.

In response to this growing metrological vulnerability, the scientific community undertook a decades-long effort to redefine the kilogram using immutable natural constants. This effort culminated in 2019 with the adoption of a new definition of the kilogram based on the Planck constant (h), a fundamental quantity in quantum physics. This redefinition was made feasible by the development and refinement of the Kibble balance, an instrument capable of relating mechanical power (*via* mass) to electromagnetic energy, enabling the measurement of mass without referencing a physical object. By fixing the numerical value of the Planck constant and redefining the kilogram accordingly, mass was fundamentally anchored to quantum phenomena rather than to an artifact subject to the imperfections of the physical world.

The transition marked a profound evolution in the philosophy and practice of metrology. It not only solved the problem of artifact drift but also aligned mass measurement with a broader scientific trend: anchoring the International System of Units (SI) in invariant constants of nature. The new definition ensures that mass standards are stable, reproducible, and universally accessible across all times and places, benefiting laboratories, industries, and researchers alike. This shift also reflects a maturing of metrology into a discipline that now fully embraces quantum-based measurement science, laying the groundwork for future advances in fundamental physics and precision manufacturing.

Root Cause

The root cause of this drift was inherent to the system: a physical artifact, however protected, is vulnerable to microscopic surface contamination, handling, cleaning, or material loss through evaporation or abrasion. Despite the IPK's noble alloy and controlled environment, its mass could not be guaranteed as invariant. The changes may have stemmed from surface cleaning techniques, environmental exposure, or undetected surface sorption. Compounding the problem, there was no absolute reference against which to detect whether the IPK was losing or its copies were gaining mass. The entire system was relative. Moreover, the concept of defining a fundamental SI unit by a single object contradicted the modern scientific ideal of universality and reproducibility, a core principle of metrology. The drift may have been minor in scale, but its epistemological and practical implications were enormous.

Root Cause Summary: Drift of the International Prototype Kilogram

Factor	Description
Physical artifact vulnerability	IPK's mass was subject to change due to contamination, cleaning, or material loss
Lack of absolute reference	All comparisons were relative; there was no way to confirm if IPK or replicas were drifting

(Continued)

(*Continued*)

Factor	Description
Metrological fragility	A single object served as the global anchor for all mass measurements
Surface chemistry effects	Long-term exposure to air or handling introduced micro-level changes
Scientific obsolescence	Artifact-based definitions contradicted the pursuit of constant-based units

Corrective Action

The international metrology community, led by the BIPM, NIST, and other national institutes, embarked on a decades-long effort to redefine the kilogram in terms of a fundamental physical constant. The breakthrough came through the Kibble balance (formerly Watt balance), an instrument that measures Planck's constant by equating mechanical and electrical power using quantum electrical standards. By precisely fixing the value of the Planck constant ($h = 6.62607015 \times 10^{-34}$ J·s), the kilogram could be defined independently of any material object (BIPM, 2019). After rigorous international testing, consensus was reached, and on May 20, 2019, the kilogram was officially redefined based on Planck's constant, joining the meter (defined by the speed of light) and the second (defined by atomic transitions) in a trio of constant-based SI units.

This historic shift eliminated the reliance on the IPK and removed the risk of artifact-based drift from the world's mass calibration infrastructure. National metrology laboratories now maintain mass standards through realization experiments, such as Kibble balances or silicon sphere Avogadro projects, linked to immutable constants rather than changeable materials. The redefinition not only restored trust in the stability of the kilogram, but it also reinforced a paradigm shift: that the most reliable standards in science are no longer physical; they are conceptual, constant, and universally accessible.

Corrective Action Summary: Redefinition of the Kilogram

Area	Corrective action
Redefinition of SI unit	Kilogram redefined *via* the Planck constant, replacing the artifact-based definition
Adoption of quantum standards	Use of Kibble balances and silicon spheres to realize mass from physical constants
Global consensus process	International coordination through the BIPM and Consultative Committee for Mass and Related Quantities (CCM)
Elimination of artifact drift	No longer dependent on a physical object prone to handling or surface changes
Improved stability and access	Constant-based units are universally reproducible and immune to material degradation

References

- BIPM, *On the Revision of the International System of Units (SI)*, 2018, https://www.bipm.org/en/measurement-units/rev-si.
- BIPM, *The International System of Units (SI)*, 9th, 2019.
- Robinson, I.A. and Schlamminger, S., The Watt or Kibble Balance: A Technique for Implementing the New SI Definition of the Kilogram. *Metrologia*, 53, 5, A46–A74, 2016.
- Stock, M., The Re-definition of the SI Units. *Philos. Trans. R. Soc. A*, 377, 2157, 2019.
- NIST, *Redefining the Kilogram with Planck's Constant*, 2019, https://www.nist.gov/si-redefinition/kilogram-redefinition.

Gas Chromatograph (GC) Flame Ionization Detector (FID) Calibration Drift

Industry: Petrochemical/Environmental Testing

- Timeframe: 1990s–2000s
- Issue: Long-term drift in FID response due to column degradation, temperature control errors, and aging hydrogen-air ratio settings.

- Impact: Hydrocarbon concentrations in air and fuel samples misreported; false positives or regulatory compliance errors.
- Resolution: Introduction of routine detector calibration with certified reference gases, plus software-based baseline tracking.

In the 1990s through the early 2000s, the petrochemical and environmental testing industries faced a growing concern related to calibration drift in Flame Ionization Detectors (FID) used in gas chromatography (GC) systems. The FID, a widely used detector for quantifying hydrocarbons, became prone to long-term signal drift, undermining the accuracy of measurements critical to both product quality and regulatory compliance.

The drift stemmed from several interrelated issues. First, column degradation over time due to contamination, bleeding, or physical wear altered the chromatographic response and retention characteristics, impacting detector performance. Second, temperature control inconsistencies, particularly in older GC systems, could shift retention times and peak responses, complicating quantification. Most critically, changes in the hydrogen-to-air flame ratio as components aged or were improperly maintained led to fluctuations in ionization efficiency, causing gradual but significant drift in response sensitivity.

The impact of these calibration issues was widespread. In the environmental sector, airborne hydrocarbon concentrations were sometimes under- or overreported, leading to false positives or missed violations of regulatory thresholds. In the petrochemical industry, errors in fuel sample analysis could compromise process control, quality assurance, and compliance with fuel formulation standards.

The resolution came through a combination of enhanced calibration protocols and software innovations. Laboratories and industrial users began implementing routine detector calibration using certified reference gas mixtures, ensuring consistent and traceable detector response over time. These gases allowed for quantitative validation of detector performance at known concentration levels, which helped account for any system drift.

Additionally, modern GC software introduced baseline tracking and correction algorithms, which allowed the system to monitor and adjust for detector baseline changes in real time, further improving stability and measurement confidence. Together, these advances significantly reduced the risk of undetected FID drift and helped ensure greater reliability and regulatory confidence in GC-based hydrocarbon analysis.

Between the 1990s and early 2000s, laboratories and industrial facilities that relied on gas chromatography (GC) with flame ionization detection (FID) began encountering persistent issues related to long-term calibration drift. The FID, which had long been a cornerstone of hydrocarbon analysis due to its high sensitivity and linear response to organic compounds, began to display inconsistencies that undermined data accuracy in both environmental monitoring and petrochemical quality control. As awareness of the drift problem grew, it became clear that this was not an isolated instrumentation issue but rather a systemic challenge influenced by the complex interplay of detector maintenance, system aging, and analytical protocol rigor.

The causes of FID drift were multifactorial. A major contributor was column degradation, a natural consequence of prolonged use, exposure to contaminants, and the thermal stresses associated with repeated temperature cycling. As GC columns aged, their ability to separate compounds with precision declined, leading to peak tailing, retention time shifts, and decreased resolution. This degradation directly affected how the detector interpreted the eluting compounds. In parallel, temperature control irregularities in aging or poorly maintained GC ovens exacerbated these shifts. Because GC relies on stable temperature gradients to produce repeatable separations, even minor variations could distort peak shapes and amplitudes, influencing quantification.

However, the most impactful and often overlooked factor was the hydrogen-to-air ratio used to fuel the FID flame. The ionization process in the detector depends critically on maintaining an optimal balance between hydrogen, air, and sample gases. Over time, as mass flow controllers aged or were recalibrated inconsistently, this ratio could drift, changing the flame's stability and ionization efficiency. As a result, the same sample could produce different detector responses depending on the state of the flame, even if the column performance remained unchanged. This phenomenon was particularly insidious because it often occurred gradually, escaping immediate detection until measurement discrepancies became too large to ignore.

The consequences of this drift varied by sector but were uniformly problematic. In environmental testing, laboratories responsible for measuring ambient hydrocarbon concentrations occasionally reported false positives or failed to detect exceedances of air quality standards, creating compliance risks. In petrochemical laboratories, misreported concentrations of hydrocarbons in fuel blends or process streams led to product inconsistencies, regulatory non-conformance, and, in some cases, costly rework or fines. These errors undermined confidence in laboratory data and placed greater scrutiny on analytical procedures during audits and inspections.

The industry responded by modernizing both calibration practices and data acquisition technologies. A key development was the routine use of certified reference gas mixtures for detector calibration. These high-purity, traceable standards enabled analysts to verify detector linearity and sensitivity at known concentration levels, helping to correct or identify drift early. Equally important were the software improvements; GC systems began incorporating baseline tracking algorithms, which continuously monitored detector output and applied corrections to compensate for background signal shifts. These tools not only increased confidence in analytical results but also enhanced quality system defensibility by maintaining comprehensive digital records of calibration and detector behavior.

Together, these advancements addressed the primary vulnerabilities associated with FID calibration drift. They elevated the standard for hydrocarbon measurement accuracy in both environmental and industrial contexts and reinforced the importance of lifecycle awareness and preventive maintenance in analytical metrology. The case of FID drift remains a teaching moment for laboratories worldwide, demonstrating that even mature technologies require continuous oversight, traceable calibration practices, and adaptation to emerging operational challenges.

Root Cause

The root causes of this drift were multifaceted and often cumulative. First, GC columns gradually degraded over time, especially when exposed to high-boiling hydrocarbons, halogenated compounds, or oxidizing contaminants. This affected retention times and peak resolution, which in turn influenced detector response stability. Second, temperature control systems for both the oven and the detector cell exhibited small but compounding errors due to faulty thermocouple calibration or controller drift. Even slight temperature fluctuations altered peak area and response factor calculations. Third, the FID's performance is highly sensitive to the stoichiometric balance of hydrogen and air, which form the flame responsible for ionizing organic compounds. Aging regulators, flow controllers, or pressure valves introduced variability in this ratio over time, causing unpredictable changes in signal intensity. Finally, infrequent calibration or reliance on outdated calibration curves created a false sense of instrument stability, especially in labs operating under high throughput and tight timelines.

Root Cause Summary: GC-FID Calibration Drift

Factor	Description
Column degradation	Loss of performance due to residue buildup, phase bleeding, or thermal damage
Temperature control drift	Thermocouple or PID controller inaccuracies affect detector and retention behavior
Flame stoichiometry variability	Aging regulators or clogged lines affect the hydrogen/air ratio, altering the sensitivity
Outdated calibration curves	Prolonged reliance on historical response factors without routine recalibration
High sample throughput stress	Continuous operation without preventive maintenance exacerbates all failure modes

Corrective Action

The instrument manufacturers and regulatory bodies implemented several corrective actions that are now considered standard best practices. Chief among them is the routine use of certified calibration gas mixtures, traceable to NIST or equivalent national standards, to verify detector response on a daily or per-batch basis. Calibration checks are now often integrated into automated GC startup sequences, with software-enforced acceptance criteria for retention time windows and detector linearity. In parallel, GC software evolved to include baseline tracking algorithms and drift compensation routines, which monitor and correct for minor shifts in baseline signal between runs.

Laboratories also adopted preventive maintenance schedules, mandating regular column replacement, septum changes, and flame jet cleaning to stabilize FID performance. Enhanced documentation of calibration records linked to sample batches and instrument IDs was introduced to improve traceability and audit readiness. In some high-throughput environments, dual-column systems were implemented to cross-check results and flag calibration inconsistencies before data release. These changes restored analytical confidence and ensured that regulatory decisions based on hydrocarbon quantification were grounded in robust, traceable, and reproducible data.

Corrective Action Summary: GC-FID Calibration Reforms

Area	Corrective action
Calibration practices	Use of NIST-traceable certified reference gases for routine detector verification
Instrument baseline management	Software-based tracking and automated calibration alerts
Preventive maintenance	Scheduled replacement of columns, jets, and critical flow components
Flame stability controls	Regular tuning of hydrogen/air ratios; replacement of regulators and flow restrictors
Data integrity and traceability	Calibration logs linked to sample batches and regulatory reporting systems

References

- ASTM D5504-12, *Standard Test Method for Determination of Sulfur Compounds in Natural Gas and Gaseous Fuels by Gas Chromatography and FID*, 2012.
- U.S. Environmental Protection Agency (EPA), *Method TO-14A: Determination of Volatile Organic Compounds (VOCs) in Ambient Air Using GC-FID*, 2003.
- Bouchard, M. and Purcell, R., Sources and Detection of Drift in GC-FID Systems. *J. Chromatogr. Sci.*, 39, 11, 453–460, 2001.
- NIST, *Certified Reference Materials for Hydrocarbon Calibration Standards*, 2008.
- International Union of Pure and Applied Chemistry (IUPAC), *Recommendations on Flame Ionization Detection Sensitivity and Calibration Protocols*, 1997.

Optical Emission Spectrometer (OES) Metal Analysis Drift

Industry: Metallurgy/Foundries

- Timeframe: Ongoing risk
- Issue: Spectrometers slowly drift due to lamp degradation, plasma chamber contamination, and matrix effect changes (composition-specific interference).

- Impact: Alloy certification errors (e.g., mislabeling 316 vs. 304 stainless steel), product recalls, and safety risks.
- Safeguard: Standardization of multi-point calibration using certified metal blocks and frequent recalibration checks against traceable standards.

In the metallurgy and foundry industries, optical emission spectrometry (OES) remains a cornerstone technology for rapid elemental analysis of metal alloys. However, despite its precision and speed, OES instruments face an ongoing risk of calibration drift, which can compromise the accuracy of alloy identification and compositional certification.

Several factors contribute to this drift. Over time, degradation of the excitation source (typically a spark or arc lamp) leads to reduced intensity or inconsistent energy delivery to the sample, altering spectral emission profiles. Additionally, contamination or wear in the plasma chamber, such as a buildup of oxide residues or sputtered metal particles, can affect signal quality and introduce spectral noise. A further challenge comes from matrix effects, inter-element interference, or changes in the sample's base composition that shift peak intensities, especially in complex or recycled alloys.

The impact of uncorrected OES drift can be severe. It may lead to alloy misclassification, such as confusing 316 stainless steel for 304, which can result in product recalls, quality failures, or safety incidents in industries ranging from construction to medical device manufacturing. In critical applications such as aerospace, pressure vessels, or chemical processing, incorrect alloy certification due to OES drift can pose serious reliability and liability risks.

To address this, the industry has adopted rigorous safeguards, particularly the standardization of multi-point calibration routines using certified reference metal blocks that represent a broad compositional range. These calibration standards are traceable to national metrology institutes and are used to establish and verify the instrument's response across various alloy matrices.

Moreover, frequent recalibration checks often performed at the start of each shift or batch ensure that any drift is quickly detected and corrected. Some systems also integrate software-based drift compensation and standardization protocols, comparing real-time spectral data against baseline profiles. Together, these practices maintain high analytical confidence and safeguard the integrity of metal certification processes across global supply chains.

In the metallurgy and foundry industries, optical emission spectrometry (OES) plays an essential role in the rapid, on-site analysis of metallic elements, enabling real-time decisions about alloy composition, melt control, and product certification. Despite its widespread use and robust analytical capabilities, OES systems are susceptible to calibration drift, a persistent challenge that threatens the integrity of compositional data over time. This drift, if not managed proactively, can lead to critical misclassifications of metal alloys, such as mistaking 316 stainless steel for 304, triggering costly consequences ranging from product recalls to safety incidents.

The drift in OES performance arises from a combination of mechanical and chemical phenomena that unfold gradually during routine operation. A major contributor is the degradation of the excitation source, typically a high-voltage spark or arc that excites atoms in the metal sample to emit characteristic light spectra. As the excitation system ages, especially the electrodes or spark source components, the energy transfer becomes less consistent, leading to fluctuations in the emitted spectra. Another persistent issue is plasma chamber contamination, which occurs as fine oxide particles or sputtered metal from repeated testing accumulate on optical components or within the sample stand. These residues scatter or absorb light, subtly distorting spectral intensities and reducing the signal-to-noise ratio.

A more complex contributor to drift is the matrix effect, in which the presence and ratios of base metals and alloying elements influence the behavior of spectral lines. In recycled or highly variable alloys, these interactions can shift the apparent intensities of emission lines, especially when compared to cleaner, more homogeneous calibration standards. The result is a creeping distortion of the calibration curve, which, if not corrected, causes the instrument to report incorrect concentrations for key elements like chromium, nickel, or molybdenum. This type of error is especially dangerous in industries that rely on precise alloy specifications for corrosion resistance, strength, or regulatory compliance.

The potential impact of OES drift extends far beyond the laboratory. Misclassification of metal alloys can result in products being deployed with incorrect mechanical or chemical properties, potentially compromising structural integrity or process compatibility. In sectors such as aerospace, energy, and biomedical manufacturing, where materials must meet strict standards, a single misidentified heat of metal can trigger a cascade of quality issues, regulatory violations, and liability concerns. Moreover, in an era of increasingly globalized supply chains, an undetected drift event in one facility can propagate through multiple tiers of manufacturing before it is discovered, amplifying its cost and complexity.

To mitigate these risks, the industry has implemented a series of rigorous safeguards. One of the most effective strategies is the use of multi-point calibration procedures that involve analyzing certified reference metal blocks representing a range of alloy compositions. These blocks are traceable to national metrology institutes, providing a reliable foundation for calibrating the spectrometer's response to different element concentrations. Frequent recalibration checks are also standard practice, often performed at the beginning of each production shift, analytical batch, or melt analysis. These checks detect early signs of drift and allow for timely correction before production decisions are made based on faulty data.

Advanced OES systems further support drift mitigation through software-based standardization protocols that continuously compare live spectra against stored baseline profiles. When deviations are detected, automated adjustments can be applied, or operator alerts can be triggered to initiate manual recalibration. Together, these technologies and procedural controls ensure that OES continues to provide the high level of analytical confidence required for critical metal certification tasks. The case of OES drift serves as a compelling example of why even powerful analytical technologies must be supported by strong metrological infrastructure and routine calibration discipline to maintain their reliability in the field.

Root Cause

The root causes of OES drift are both mechanical and analytical. Over time, the spectral source components, including excitation lamps, spark electrodes, and plasma chambers, degrade due to heat cycling, material buildup, and atmospheric contamination. This alters the consistency of the plasma and signal intensity at specific wavelengths. Equally problematic is the accumulation of metallic residues and oxide films on the sample stand and electrode tip, which affect arc stability and spectral reproducibility. Another major contributor is the matrix effect, in which minor compositional differences (such as varying base metal or impurity levels) influence the ionization and excitation environment, skewing the analytical response for specific elements. Many systems also suffer from software oversimplification, where previous calibration curves are used without verification that the detector response still matches real-world standards, particularly after power interruptions or hardware maintenance. In high-throughput operations, this cumulative drift may go unnoticed until audit failure or a downstream metallurgical test reveals the discrepancy.

Root Cause Summary: OES Spectrometer Drift in Alloy Analysis

Factor	Description
Lamp and electrode aging	Excitation sources degrade, reducing plasma consistency and line intensity
Plasma chamber contamination	Residue and oxidation alter arc behavior and emission reproducibility
Matrix effect shifts	Base metal or impurity variations distort element-specific calibration
Calibration curve staleness	Calibration drift occurs without recalibration or the use of outdated standard curves
Maintenance oversight	Accumulated drift unnoticed due to infrequent or informal recalibration

Corrective Action

The **routine use of certified reference materials (CRMs)** typically involves **multi-point calibration blocks** that span the composition range of alloys in production. These CRMs are sourced from accredited bodies (e.g., NIST, BAM, or BCR) and are matched to the expected matrix type (e.g., austenitic stainless, low-alloy steel, or aluminum-silicon). Analysts now perform **daily verification and drift correction routines**, comparing internal results with CRM benchmarks to detect deviations. Modern OES software suites support **automated recalibration prompts**, error thresholds, and **intensity ratio diagnostics**, which alert users to degraded signal behavior across multiple emission lines.

Additionally, procedural reforms emphasize **instrument maintenance routines**, including cleaning of electrodes and plasma stands, reconditioning of sample surfaces, and **periodic factory recalibration of the detector optics**. Instrument manufacturers now recommend, and many quality assurance standards require, **traceable calibration logs** linked to production batches, enabling full accountability and recall traceability. In advanced applications, such as aerospace or nuclear-grade metallurgy, dual-path verification using **portable XRF** or **wet chemical analysis** is occasionally mandated to validate critical OES readings. These safeguards, when consistently applied, have restored confidence in OES as a fast, reliable, and quantifiable method of alloy certification while reinforcing the principle that **fast does not excuse calibration complacency**.

Corrective Action Summary: OES Drift Mitigation

Area	Corrective action
Calibration verification	Frequent checks using certified multi-element metal blocks across relevant matrix types
Matrix-matched standards	Use of CRMs that reflect the exact alloy class under test
Software-driven recalibration	Auto-prompts for calibration based on usage, signal behavior, or drift detection
Maintenance protocols	Routine electrode cleaning, plasma stand servicing, and optics recalibration
Data traceability	Calibration logs tied to batch reports for audit and recall traceability

References

- ASTM E415-17, *Standard Test Method for Analysis of Carbon and Low-Alloy Steel by Spark Atomic Emission Spectrometry*, 2017.
- ISO 17025:2017, *General Requirements for the Competence of Testing and Calibration Laboratories.*
- NIST SRM Catalog, *Standard Reference Materials for Metal Alloys and Spectroscopic Calibration*, Ongoing.
- Boehm, P. and Valente, A., Mitigation of Drift in Spark-OES Systems: Best Practices and Design Considerations. *J. Anal. At. Spectrom.*, 18, 4, 361–366, 2003.
- BIPM, *International Vocabulary of Metrology – Basic and General Concepts and Associated Terms (VIM)*, 2020.

Industrial Scales – Load Cell Drift

Industry: Logistics/Pharmaceuticals/Food Processing

- Timeframe: 2000s–present
- Issue: Load cells in industrial and laboratory balances showed long-term drift due to temperature, creep, and electrical noise.
- Impact: Product fill errors, dosage mislabeling, regulatory audit failures.

- Resolution: ISO/IEC 17025 calibration intervals, with automatic drift compensation and logging in modern balances.

In industries where precision mass measurement is critical, such as logistics, pharmaceuticals, and food processing, the long-term drift of load cells in industrial and laboratory balances has become a persistent and impactful issue since the early 2000s. Load cells, which convert mechanical force into electrical signals, are inherently susceptible to drift over time due to a combination of temperature fluctuations, mechanical creep, moisture ingress, and electrical noise.

This drift often occurs gradually and invisibly, making it difficult to detect without routine verification. As the load cell's output signal drifts away from its true value, the scale begins to under- or over-report mass measurements, potentially without triggering any error warnings. In high-throughput operations or critical dosing environments, this can have serious consequences.

The impact of undetected load cell drift includes product fill inaccuracies, which may result in underfilled packages or excessive overfilling, both of which carry cost and compliance risks. In the pharmaceutical industry, even small deviations can lead to dosage mislabeling or formulation errors, violating strict regulatory standards and posing patient safety risks. In the food sector, inconsistent fill weights can lead to labeling violations and audit failures, especially in jurisdictions with tight consumer protection laws.

To mitigate these risks, modern safeguards include adherence to ISO/IEC 17025 calibration requirements, which mandate regular calibration intervals using traceable mass standards. These protocols ensure that balances remain within specification and that any drift is promptly identified and corrected.

In addition, modern industrial and laboratory balances are now equipped with automatic drift compensation algorithms, temperature stabilization, and built-in diagnostics that can detect and alert users to abnormal deviations. These systems often log calibration history and measurement drift trends, enabling traceability during regulatory audits and supporting continuous improvement in quality assurance practices. This combination of scheduled recalibration and intelligent monitoring has significantly improved the reliability and compliance of mass measurement systems in demanding industrial settings.

Since the early 2000s, the issue of load cell drift in industrial and laboratory balances has emerged as a significant challenge in sectors where precise mass measurement is non-negotiable. Industries such as logistics, pharmaceuticals, and food processing rely on accurate weighing systems for everything from product formulation and fill weight compliance to inventory tracking and quality control. Central to these systems are load cells and electromechanical transducers that convert applied mechanical force into a measurable electrical signal. Despite their precision, load cells are susceptible to gradual and cumulative drift, driven by environmental and operational factors that affect the reliability of the measurements they produce.

The drift typically develops over time due to several interconnected factors. Temperature fluctuations cause expansion and contraction in the load cell material, altering the strain gauge response. Mechanical creep, or the tendency of materials to deform permanently under prolonged stress, can also shift the zero point or sensitivity of the cell. Additional sources of drift include moisture ingress, which may interfere with electrical insulation or corrode internal circuitry, and electrical noise or interference, particularly in industrial environments with high electromagnetic activity. Because the changes tend to be incremental, operators may not notice them until they cause substantial discrepancies, often after many production cycles have passed with inaccurate measurements.

The consequences of undetected load cell drift can be far-reaching. In high-throughput logistics operations, weight inaccuracies may lead to incorrect inventory counts or shipping cost errors. In the pharmaceutical industry, even slight deviations can cause incorrect dosage labeling or formulation inconsistencies, posing direct risks to patient safety and violating stringent regulatory frameworks such as those enforced by the FDA or EMA. In food processing, inaccurate fill weights can result in packages that do not meet declared weight standards, leading to consumer complaints, recalls, or penalties under fair packaging and labeling laws. More critically, such deviations can be discovered during a regulatory audit, potentially triggering corrective actions, production shutdowns, or legal liabilities.

To address these risks, industry standards and best practices have evolved significantly. The adoption of ISO/IEC 17025 calibration protocols has been a cornerstone in formalizing how and when balances should be calibrated. These standards require routine calibration intervals using traceable reference weights and documented procedures that ensure consistency, repeatability, and legal defensibility. Furthermore, manufacturers have responded by incorporating automated drift compensation features in modern balances. These features include temperature stabilization algorithms, self-diagnostic tools, and real-time deviation monitoring, all designed to detect anomalies before they result in nonconforming product output.

Additionally, modern weighing systems are equipped with digital logging capabilities, allowing facilities to maintain detailed records of calibration history, detected drift patterns, and system alerts. This traceability not only is essential for regulatory compliance but also provides a foundation for continuous improvement in quality management systems. Together, these advances in hardware, software, and procedural oversight have significantly reduced the risks associated with load cell drift, enhancing the confidence with which mass measurements are used in critical industrial and laboratory operations.

Root Cause

The root cause of load cell drift lies in both physical and electrical vulnerabilities. First, temperature effects, particularly in unregulated or fluctuating environments, can cause the strain gauge material to expand or contract, leading to zero-point shifts and apparent weight drift. Second, material creep, especially in lower-cost load cell alloys, causes permanent deformation over time under constant load, subtly altering the system's mechanical baseline. Third, electrical noise and grounding issues, particularly in facilities with nearby high-power equipment or variable frequency drives, introduce erratic or biased signals into the readout. Additionally, infrequent recalibration or overreliance on initial factory certification allowed gradual deviations to persist undetected. In industries like pharmaceuticals, where gravimetric accuracy is linked directly to product safety and labeling compliance, these failures had amplified implications.

Root Cause Summary: Load Cell Drift in Industrial and Laboratory Balances

Factor	Description
Thermal expansion and sensitivity	Load cells respond to ambient temperature shifts, causing zero or span drift
Mechanical creep	Materials under long-term load deform, altering baseline weight calibration

(Continued)

(*Continued*)

Factor	Description
Electrical interference	EMI and grounding issues distort strain gauge signals and introduce instability
Lack of periodic recalibration	Systems were left unchecked for months or years, masking cumulative drift
Overreliance on factory certs	Initial calibration assumed stable without adjustment for local conditions

Corrective Action

Corrective actions implemented a suite grounded in standards and automation. Foremost among these is the application of ISO/IEC 17025:2017, which defines the requirements for competent calibration laboratories and mandates routine recalibration intervals, especially for weighing instruments used in regulated industries. Equipment manufacturers now design balances with automatic drift compensation algorithms, which use internal diagnostics to detect zero shifts or inconsistent readings and either correct them algorithmically or flag the system for service. Additionally, most modern scales now maintain digital calibration and drift logs, enabling traceable performance history for regulatory inspections and preventive maintenance.

Environmental safeguards have also become standard practice: high-precision scales are now routinely installed in temperature-controlled, vibration-isolated enclosures to minimize environmental influence. In sensitive industries like pharmaceutical manufacturing and high-speed packaging, automated checkweighers with in-line calibration checks have become the norm. These systems compare actual weight readings to stored reference standards during production and automatically trigger recalibration, alarm conditions, or removal from service if deviations exceed accepted tolerances. Through these reforms, organizations have significantly reduced the risk of undetected weighing errors, reinforcing the principle that precision without verification is an illusion.

Corrective Action Summary: Mitigating Load Cell Drift in Industrial Scales

Area	Corrective action
Calibration standards	ISO/IEC 17025-compliant recalibration intervals and procedures
Automated drift compensation	Built-in firmware algorithms for zero tracking and span correction
Environmental controls	Temperature- and vibration-controlled environments for high-precision balances
In-line calibration monitoring	Use of checkweighers and calibration routines during real-time operation
Digital traceability	Logging of drift, calibration history, and user interactions for audit readiness

References

- ISO/IEC 17025:2017, *General Requirements for the Competence of Testing and Calibration Laboratories.*
- OIML R76-1, *Non-Automatic Weighing Instruments – Requirements and Test Procedures*, 2006.
- NIST Handbook 44, *Specifications, Tolerances, and Other Technical Requirements for Weighing Devices*, 2023.
- US Pharmacopeia (USP) <41> and <1251>, *Balances and Weighing on an Analytical Balance.*
- Sikorski, J. and Stobiecka, K., Mitigation of Drift in Load Cell Systems under Dynamic Conditions. *Meas. Sci. Technol.*, 26, 12, 125102, 2015.

Spectrophotometer Baseline Shift in Color Measurement

Industry: Paint, Plastics, Textiles

- Timeframe: Ongoing
- Issue: Changes in lamp output or detector sensitivity caused baseline shifts in CIE Lab values used to judge color quality.

- Impact: Off-spec color batches despite raw material consistency.
- Safeguard: Daily verification with certified color tiles and dual-beam spectrophotometer designs to cancel drift.

In industries where color accuracy is mission-critical, such as paint, plastics, and textiles, the reliability of spectrophotometric color measurement has long been challenged by the issue of baseline shift. This phenomenon, still an ongoing concern, is caused primarily by lamp aging, changes in detector sensitivity, and optical component degradation, all of which affect the instrument's ability to maintain a consistent baseline for reflectance or transmittance readings.

These baseline shifts can result in measurable deviations in CIE Lab values, the standardized color space used to quantify and control product color. What makes this especially problematic is that such shifts can mimic color variation, even when raw materials and production conditions remain consistent. The result: off-spec batches, rejected shipments, and costly rework not because the product actually changed, but because the measurement system did.

The impact extends beyond quality control. In tightly controlled supply chains such as automotive coatings or branded textile lines, a color out of tolerance can disrupt production, damage brand consistency, and erode customer trust. Even minor shifts ($\Delta E > 1$) can be critical when color matching is essential to product identity.

To combat this, the industry has implemented robust safeguards. One key practice is daily verification using certified color tiles, which provide a stable, traceable reference against which the instrument's performance can be checked before production measurements are taken. These tiles are calibrated to international color standards and allow users to detect baseline drift early.

Additionally, dual-beam spectrophotometer designs are now widely used. These instruments measure the sample and a reference beam simultaneously, enabling them to actively cancel out fluctuations in lamp output or detector response, significantly improving measurement stability over time.

Together, these measures ensure greater consistency in color quality, allowing manufacturers to detect true process changes rather than instrument-induced anomalies. They also reinforce the broader principle that instrument calibration and validation are just as critical as material control in delivering high-quality, color-consistent products.

In the paint, plastics, and textile industries, where visual uniformity and precise color matching are vital to product integrity and brand identity, the performance of spectrophotometers plays a pivotal role in ensuring color quality. However, one persistent challenge undermining this assurance is the baseline shift in color measurement instruments. This phenomenon, characterized by subtle but measurable changes in the spectrophotometer's baseline reflectance or transmittance values, can distort CIE Lab color readings even when the actual product has not changed. It remains an ongoing issue in production environments where environmental stresses, component wear, and calibration drift gradually accumulate.

The root of the problem lies in the aging of internal components, especially the light source (typically a xenon or tungsten-halogen lamp), which over time loses intensity or spectral uniformity. Similarly, photodetectors, particularly in the visible and near-infrared ranges, may degrade due to exposure to light, temperature variation, or electronic noise. These shifts affect the instrument's reference calibration and ultimately lead to deviation in CIE Lab values, which are used universally to express and control color. The concerning part is that these baseline drifts can simulate color variation in otherwise identical products, misleading quality control systems and operators into rejecting batches that are, in fact, within specification.

The consequences of this are more than just technical errors; they are operational and financial. Color discrepancies, even with a ΔE (Delta E) value greater than 1, can result in batch rejections, expensive rework, or disputes with customers and brand owners. In tightly controlled industries such as automotive coatings, cosmetics, branded apparel, and plastics for consumer goods, color consistency is a hallmark of quality. Even minor deviations can disrupt production schedules, delay product launches, and damage brand perception. These issues often trace back not to formulation errors, but to undetected instrument drift.

To mitigate the risk of baseline shift, the industry has adopted robust countermeasures. One key safeguard is the daily verification of instrument accuracy using certified color tiles. These tiles are precisely manufactured and calibrated to international standards such as those maintained by the National Institute of Standards and Technology (NIST). By measuring the reflectance of these known standards before production runs, operators can immediately detect shifts in baseline values and halt measurements until the instrument is recalibrated or serviced. This practice embeds traceability into routine operations and enables early intervention before bad data drives bad decisions.

Further enhancing stability, many manufacturers have transitioned to dual-beam spectrophotometer designs, which provide significant improvements in measurement consistency. In these systems, the instrument splits the light path so that one beam passes through the sample while the other travels a reference path. By simultaneously measuring both beams, the spectrophotometer can actively compensate for fluctuations in lamp output and detector sensitivity, effectively canceling out drift and extending calibration intervals. This architecture greatly reduces the chances of baseline shift impacting measurement reliability, making it especially suitable for high-volume, high-specification environments.

These practices underscore a larger truth in modern manufacturing: instrument performance is as critical as process control. Even the most consistent raw materials and stable production lines cannot ensure quality if the measurement systems themselves are compromised. In color-sensitive industries, rigorous instrument validation, including both mechanical calibration and optical verification, is essential for maintaining trust in the data and delivering products that meet the tightest visual specifications.

Root Cause

The root cause of this issue lies in the physical and optical components of the spectrophotometer. Over time, lamps degrade in intensity and spectral output, altering the amount and distribution of light reaching the sample and reference detectors. Similarly, photodetectors or photodiodes age or become contaminated, especially under heavy use or poor maintenance, leading to inconsistencies in signal detection. This results in baseline shifts, where the measured reflectance curve deviates slightly across all wavelengths, subtly but significantly altering computed CIE Lab values, especially in sensitive a^* and b^* channels that measure chromaticity. These shifts can push a color just beyond acceptable tolerance thresholds (e.g., $\Delta E > 1.0$), triggering batch rejection. Compounding the problem, some older spectrophotometers lacked self-calibration routines or failed to detect gradual performance loss, especially in single-beam instruments with no real-time correction for light source drift or environmental fluctuations.

Root Cause Summary: Spectrophotometer Baseline Shift in Color Measurement

Factor	Description
Lamp aging and intensity drift	Light source output changes over time, altering the spectral balance
Detector sensitivity loss	Aging or contaminated photodetectors change signal response at specific wavelengths
Environmental instability	Temperature, humidity, or dust cause subtle optical shifts
Single-beam instrument limitations	Lack of an internal reference beam prevents real-time drift correction
Infrequent verification	Baseline shifts go unnoticed without regular checks using certified color standards

Correct Action

Manufacturers and quality laboratories now employ a range of **corrective actions and safeguards**. The most impactful has been the **routine use of certified color standards**, such as **ceramic or glass calibration tiles** with traceable CIE Lab values. These tiles are used daily or before each measurement session to verify that the spectrophotometer is producing accurate baseline readings. If deviations exceed predefined tolerances, recalibration is triggered automatically or manually before production measurements resume. Additionally, modern spectrophotometers are increasingly based on **dual-beam optical architectures**, which simultaneously monitor the sample and reference paths, enabling real-time compensation for lamp or detector drift.

Instrument software has also advanced, incorporating **automated baseline correction algorithms, internal diagnostics**, and **timed maintenance reminders** for lamp replacement and detector cleaning. Furthermore, **digital calibration logs** are now standard in quality systems, ensuring that any deviation from standard values is traceable and auditable. For industries requiring tight color tolerances, instruments are periodically requalified using **traceable colorimetric standards from NIST, BCRA, or equivalent bodies**. These changes have strengthened the role of the spectrophotometer from a basic quality control tool into a **fully validated, calibration-dependent instrument**, critical to consistent product color and brand identity.

Corrective Action Summary: Managing Baseline Shift in Color Spectrophotometry

Area	Corrective action
Daily verification	Use of certified color tiles with known CIE Lab values to detect drift before measurement
Dual-beam instrument design	Simultaneous sample/reference path measurement to cancel out lamp and detector drift
Preventive maintenance	Scheduled lamp replacement, detector cleaning, and environmental monitoring
Software diagnostics	Built-in calibration routines, baseline correction, and error detection algorithms
Traceability and logging	Electronic logs of calibration checks and ΔE deviations linked to production batches

References

- ASTM E1164-12, *Standard Practice for Obtaining Spectrometric Data for Object-Color Evaluation*, 2012.
- ISO 11664-4:2008, Colorimetry – Part 4: CIE 1976 Lab Colour Space*.
- NIST Special Publication 250-48, *Standard Reference Materials for Color and Reflectance Calibration*, 2005.
- Konica Minolta, *Causes and Solutions for Baseline Drift in Spectrophotometry*, 2016.
- HunterLab, *Best Practices for Spectrophotometer Calibration and Verification*, 2020.

Flow Meter Calibration Drift in Custody Transfer

Industry: Oil & Gas/Utilities

- Timeframe: 1990s–2010s
- Issue: Orifice plate and turbine flow meters drifted due to wear, fouling, or erosion, undetected between calibrations.

- Impact: Financial losses in custody transfer (billions of dollars in some contracts).
- Resolution: Use of ultrasonic and Coriolis flow meters with onboard self-diagnostics and redundancy.

In the oil & gas and utilities sectors, where precision in custody transfer is essential for multimillion- and sometimes billion-dollar transactions, flow meter calibration drift posed a serious challenge from the 1990s through the 2010s. Traditional flow measurement technologies, particularly orifice plate and turbine flow meters, suffered from long-term drift due to wear, fouling, erosion, and mechanical degradation. These effects often went undetected between calibration intervals, especially in remote or high-throughput installations.

In orifice plate systems, plate edge wear or buildup could subtly change flow profiles, leading to under- or over-registration of volume. Turbine meters, meanwhile, experienced bearing wear, rotor imbalance, and debris interference, all of which degraded performance over time. In many cases, the meters continued to function but delivered inaccurate flow readings that went unnoticed without frequent recalibration or performance checks.

The impact was profound. In custody transfer, even small percentage errors in flow measurement can lead to major financial discrepancies, especially when dealing with high-value commodities like natural gas, crude oil, or refined fuels. Over time, these undetected drifts contributed to billing disputes, regulatory challenges, and financial losses amounting to millions or even billions of dollars across long-term contracts.

To address this, the industry underwent a technological transformation. Ultrasonic and Coriolis flow meters emerged as superior alternatives, offering non-intrusive, wear-free operation and significantly improved measurement stability. These modern systems include onboard self-diagnostic tools that continuously monitor signal integrity, flow profile, and sensor performance, enabling early detection of anomalies or drift.

Moreover, redundant meter installations and automated validation routines became standard in critical custody transfer points, ensuring continuous verification and eliminating sole reliance on a single measurement source. These safeguards drastically improved confidence in flow measurement accuracy, aligning financial accountability with technical reliability, and setting a new standard for metrological assurance in high-value transfer environments.

In the oil and gas, as well as utility industries, custody transfer flow measurement represents one of the most financially sensitive applications of metrology. Custody transfer refers to the point at which ownership of a commodity such as crude oil, natural gas, or refined fuels changes hands, often triggering multimillion-dollar transactions. From the 1990s through the 2010s, the sector increasingly recognized a serious vulnerability in its legacy flow measurement systems: long-term calibration drift in orifice plate and turbine flow meters. These conventional technologies, long considered industry staples, were prone to gradual degradation from mechanical wear, fouling, and erosion. Because these effects were often subtle and occurred between infrequent calibration intervals, they introduced undetected errors into some of the most economically consequential measurements in industrial operations.

Orifice plates, among the oldest and most widespread flow technologies, rely on a pressure differential created by a constriction in the flow path. However, edge wear or material buildup on the plate, even microscopic, can alter flow profiles and result in significant measurement errors. The situation is exacerbated in real-world environments where flow composition includes corrosive gases, suspended solids, or waxy residues that degrade the orifice geometry over time. Turbine meters, which measure flow by tracking the rotational speed of an internal rotor, are similarly susceptible to drift. Bearings wear down, rotors can become imbalanced, and foreign debris can obstruct the mechanism, all of which reduce the accuracy of readings. These devices often continue to operate and produce plausible numbers, but the output can gradually diverge from reality, undetected unless rigorous recalibration protocols are enforced.

The economic consequences of such drift are immense. A small percentage error in flow measurement, say 0.5%, may seem insignificant in isolation. However, over the course of a year, in a high-volume natural gas pipeline or offshore oil offloading terminal, that error can translate into millions or even billions of dollars in disputed value. These inaccuracies have led to regulatory investigations, audit disputes, and legal battles between operators, suppliers, and customers. In some documented cases, long-term contractual imbalances created by undetected drift resulted in substantial financial settlements and damaged reputations.

Recognizing the systemic risk posed by aging mechanical flow meters, the industry embraced a wave of technological innovation and procedural reform. The adoption of ultrasonic flow meters marked a major turning point. These meters use transit-time or Doppler techniques to measure flow velocity non-invasively, avoiding the mechanical degradation issues associated with moving parts. Similarly, Coriolis flow meters, which directly measure mass flow *via* the Coriolis effect in vibrating tubes, provided an alternative with no moving parts and inherently higher accuracy and stability. Both technologies offer self-diagnostic capabilities, monitoring signal quality, flow profile symmetry, and temperature sensitivity in real time, which enable early warning of any performance degradation.

In addition to deploying advanced metering technologies, operators began implementing redundant measurement systems and automated verification routines. At critical custody transfer points, two or more flow meters are now often installed in parallel, allowing data validation through cross-comparison. Software tools track long-term trends and alert operators to discrepancies that may indicate drift or fouling. This multi-layered approach has drastically improved confidence in flow measurement, reducing the potential for uncorrected drift to impact billing or contractual accountability.

The transition from mechanical to digital, self-monitoring flow measurement in custody transfer environments reflects a broader evolution in industrial metrology, one where instrument reliability and data transparency are essential not only for operational performance but also for financial accuracy and legal defensibility. As contracts become more data-driven and regulatory scrutiny intensifies, the lessons from historical drift incidents continue to underscore the importance of proactive calibration oversight and investment in robust, drift-resistant measurement infrastructure.

Root Cause

The root causes of these measurement errors were diverse but systematic. Orifice plates, which rely on fixed pressure differential geometry, experienced edge wear, deformation, or build-up of residue, subtly changing the flow profile and reducing the accuracy of the differential pressure calculation. Turbine meters suffered from bearing degradation, rotor wear, or debris fouling, causing under-rotation or erratic output signals. Compounding the problem was the assumption of stable meter coefficients that once-calibrated devices would maintain their performance over time. Field conditions, however, introduced temperature fluctuations, viscosity changes, multiphase flow, and pulsation, all of which further degraded accuracy. Most importantly, these meters provided no self-diagnostic feedback or performance assurance in real time. Without frequent site visits or recalibration with portable prover loops, operators had no means of detecting drift between scheduled calibrations.

Root Cause Summary: Flow Meter Drift in Custody Transfer

Factor	Description
Mechanical wear and erosion	Orifice plates and turbine blades degrade with time and flow velocity
Fouling and contamination	Residue buildup alters internal geometry and flow profile
Environmental and process variation	Temperature, viscosity, or multiphase conditions introduce error
Calibration interval gaps	Long periods between calibrations allowed drift to go unnoticed
No real-time diagnostics	Legacy meters lacked onboard feedback to alert users of accuracy degradation

Corrective Action

To solve this problem and restore confidence in high-stakes custody transfer, the industry transitioned to **next-generation flow meters**, specifically **ultrasonic and Coriolis flow meters** designed with **real-time diagnostics, non-intrusive measurement**, and **greater stability over time**. Ultrasonic meters use multiple transit-time paths to calculate flow velocity and detect anomalies such as asymmetrical profiles (from fouling or swirl). Coriolis meters, while more expensive, provide **direct mass flow measurement** and are **immune to density, viscosity, and temperature effects**, making them ideal for refined product transfer. Crucially, both technologies support **onboard self-diagnostics**, including **signal strength monitoring, path balance checks, zero-flow detection**, and drift trending over time.

In addition, calibration practices were updated to include **proving routines with traceable reference devices**, regular comparisons to **master meters**, and **remote performance validation** *via* SCADA-integrated data analysis. Contractual language evolved to require **meter validation logs, calibration drift documentation**, and **traceable audit trails** for financial accountability. Standards such as **API MPMS Chapter 5** (Metering) and **ISO 5167** (for differential flow measurement) were revised to emphasize **calibration verification frequency**, **installation effects**, and **diagnostics integration**. These reforms have reshaped custody transfer from a static, point-in-time calibration model to a **continuous performance assurance system**, greatly reducing economic risk and strengthening the integrity of commercial transactions.

Corrective Action Summary: Flow Meter Drift Mitigation in Custody Transfer

Area	Corrective action
Meter technology upgrade	Adoption of ultrasonic and Coriolis meters with higher stability and built-in diagnostics
Real-time monitoring	Implementation of self-checking algorithms and SCADA alerts for meter performance drift
Improved calibration practices	Use of mobile provers, master meters, and traceable test routines during field checks
Contractual safeguards	Inclusion of calibration and validation logs in financial agreements
Standards revision	Updates to API MPMS and ISO 5167 for traceability, installation geometry, and recalibration frequency

References

- API MPMS Chapter 5.5, *Measurement of Liquid Hydrocarbons by Coriolis Meters*, 2016.
- API MPMS Chapter 5.8, *Measurement of Liquid Hydrocarbons by Ultrasonic Flow Meters*, 2017.
- ISO 5167-2:2022, *Measurement of Fluid Flow by Means of Pressure Differential Devices – Orifice Plates.*
- OIML R117-1, *Measuring Systems for Liquids Other Than Water – General Technical Requirements*, 2019.
- van Putten, H., *et al.*, Custody Transfer Flowmetering: The Next Generation. *Flow Meas. Instrum.*, 22, 4, 328–334, 2011.
- Lipták, B.G. (Ed.), *Instrument Engineers' Handbook – Process Measurement and Analysis*, 4th, 2015.

Ion-Selective Electrode (ISE) Calibration for Sodium and Chloride in Water

Industry: Water Quality/Biomed

- Timeframe: Ongoing

- Issue: ISEs show nonlinear drift due to membrane degradation and ion interference.
- Impact: Misreading of electrolyte levels in dialysis or wastewater treatment.
- Safeguard: Calibration with fresh standard solutions and two-point verification at minimum.

In both water quality monitoring and biomedical applications such as dialysis, ion-selective electrodes (ISEs) are commonly used for measuring key electrolytes, especially sodium (Na^+) and chloride (Cl^-). However, these sensors are subject to nonlinear calibration drift, an issue that continues to present risks due to membrane degradation, ion interference, and electrode aging.

ISEs rely on selective membranes that respond to the activity of specific ions in solution, but over time, these membranes lose selectivity due to chemical wear, fouling, and ion saturation, especially in harsh or high-ionic-strength environments. Compounding the problem is interference from competing ions (e.g., potassium or bicarbonate), which can distort readings, particularly in complex matrices like wastewater or blood plasma.

The impact of drift in ISE measurements can be serious. In dialysis, incorrect sodium or chloride readings may lead to improper electrolyte balance in patients, posing immediate health risks. In wastewater treatment, inaccurate readings can disrupt process control and regulatory compliance, potentially leading to environmental discharge violations or fines.

To safeguard against these issues, best practices now require routine calibration using fresh, certified standard solutions. Critically, a minimum of two-point calibration is recommended to account for the sensor's nonlinear response curve, especially at high and low ion concentrations. Regular calibration helps track and compensate for membrane wear and sensor aging.

Additionally, electrode conditioning and frequent recalibration intervals are emphasized, along with electrode replacement schedules based on use and performance drift. Some modern analyzers also feature built-in diagnostic checks and calibration reminders, ensuring traceable and reliable operation.

These practices reinforce the importance of metrological discipline even in simple electrochemical systems, where small errors in ion measurement can have disproportionately large effects on human health and environmental integrity.

In water quality monitoring and biomedical settings such as dialysis, ion-selective electrodes (ISEs) are fundamental tools for determining concentrations of key electrolytes, particularly sodium (Na^+) and chloride (Cl^-). These sensors provide a fast and cost-effective means of analyzing ionic activity in solution and are integral to both process control and patient care. Despite their widespread use, ISEs are susceptible to nonlinear calibration drift, a problem that continues to pose challenges in critical applications. This drift is primarily driven by membrane degradation, interference from other ions, and sensor aging, all of which erode the accuracy and reliability of these systems over time.

ISEs function by generating a potential across a selective membrane that correlates with the activity of a target ion. However, the membrane's performance deteriorates as it is exposed to varying chemical loads, temperature fluctuations, and prolonged use. Over time, the membrane loses its ion-selective capabilities due to chemical wear, membrane fouling, and ion saturation, particularly in samples with high total ionic strength or contaminant loads. In environments such as blood plasma or industrial wastewater, these issues are further exacerbated by interfering ions; for example, potassium may distort sodium readings, while bicarbonate and sulfate can interfere with chloride measurements. These interferences, combined with the nonlinear behavior of ISEs at low and high concentration ranges, make the calibration process highly sensitive and increasingly unreliable if not rigorously maintained.

The consequences of ISE drift can be serious. In dialysis applications, accurate measurement of sodium and chloride is essential to maintaining electrolyte balance in patients. An error as small as a few milliequivalents per liter can result in electrolyte imbalances, which may trigger seizures, arrhythmias, or other acute complications. Similarly, in wastewater treatment, inaccurate ion concentration readings can lead to incorrect dosing of neutralizing agents, mismanagement of treatment stages, and potential non-compliance with discharge regulations. These inaccuracies not only jeopardize process efficiency but can also result in environmental penalties and reputational damage.

To mitigate these risks, industry best practices stress routine and rigorous calibration protocols. At a minimum, a two-point calibration using fresh, certified standard solutions is required to account for the nonlinear response of ISEs across the concentration spectrum. One-point calibrations are insufficient in most cases, as they cannot be corrected for slope deviations or baseline shifts. Frequent recalibration, often daily or prior to each measurement session, is essential, especially when the ISE is used in demanding or high-stakes environments. Moreover, electrode conditioning procedures, such as soaking in ion-rich solutions, and regular membrane inspection are recommended to preserve electrode performance.

Some modern analyzers now include built-in diagnostic features, such as drift monitoring, calibration reminders, and automatic validation routines. These systems support traceable calibration records and prompt users to take corrective actions when performance thresholds are breached. Additionally, planned electrode replacement intervals based on performance history or usage hours have become a standard part of quality assurance protocols in both clinical and environmental laboratories.

Overall, the challenge of ISE calibration drift underscores the broader metrological principle that even simple, routine measurements require structured oversight and verification. The perceived simplicity of ISE technology can lead to complacency, but the impact of measurement errors in this context, whether on human health or environmental safety, can be disproportionately large. Regular calibration, proper handling, and traceability are therefore not optional but essential components of any reliable electrochemical measurement program.

Root Cause

The root cause of ISE drift is primarily chemical in nature. The sensing membrane in sodium and chloride electrodes typically consists of ion-exchange materials (e.g., glass or polymeric matrices) that deteriorate over time due to leaching, biofouling, and hydrolysis. As the membrane ages, its selectivity decreases, causing cross-sensitivity to other ions, such as potassium, calcium, or nitrate. This results in nonlinear deviations from the expected concentration-to-voltage response (typically governed by the Nernst equation). Additionally, internal reference electrodes, which maintain a stable baseline potential, can degrade due to junction clogging or internal filling solution depletion. Environmental conditions such as temperature, ionic strength, and sample matrix complexity further exacerbate the drift, especially in wastewater or dialysis fluids with fluctuating compositions. Critically, ISEs do not self-correct; calibration must be actively maintained, and calibration curves must be checked for slope consistency. Yet in many field or lab settings, ISEs are used beyond their calibration range or operated without rigorous verification, leading to data inconsistency and systemic measurement bias.

Root Cause Summary: ISE Drift in Sodium and Chloride Analysis

Factor	Description
Membrane degradation	Loss of ion selectivity due to leaching, fouling, and polymer breakdown
Reference electrode instability	Potential drift from junction clogging or electrolyte depletion
Ion interference	Cross-reactivity with competing ions causes nonlinear calibration errors
Lack of multi-point calibration	Use of single-point checks fails to detect drift in slope or offset
Environmental sensitivity	Variations in temperature, matrix, and sample pH affect electrode response

Corrective Action

To mitigate these risks, laboratories and water treatment operators have implemented several corrective actions based on best practices and standard protocols. Most importantly, ISEs must be calibrated using freshly prepared or certified standard solutions that bracket the expected sample concentration. A two-point minimum calibration is now widely enforced, allowing not only zero offset verification but also validation of the electrode slope, which should ideally fall within ±2 mV of the theoretical Nernst response (typically ~59 mV/decade for monovalent ions at 25°C). For medical and regulatory labs, additional controls such as daily quality control checks with known concentrations are used to flag failing sensors or abnormal slope deviations.

Modern ISE meters incorporate calibration reminders, automatic slope evaluation, and error codes when response drift exceeds limits. In some systems, electrode aging logs are tracked digitally to ensure timely membrane replacement or reconditioning. For clinical applications, protocols from the Clinical and Laboratory Standards Institute (CLSI) and standards like ISO 15189 mandate traceability, calibration documentation, and external proficiency testing. In environmental labs, adherence to EPA-approved methods such as EPA 9211 (ISE Measurement of Chloride) is required for compliance reporting. Collectively, these practices ensure that ISE results remain traceable, repeatable, and suitable for decision-making in both health-critical and regulatory environments.

Corrective Action Summary: Managing ISE Drift for Sodium and Chloride

Area	Corrective action
Calibration verification	Use of fresh, traceable two-point standards to monitor offset and slope
Electrode lifecycle management	Routine replacement or reconditioning based on usage logs and drift thresholds
Interference testing	Verification of response in the presence of potential competing ions
Temperature and matrix control	Stabilization or compensation for temperature and ionic strength effects
Regulatory compliance	Adherence to CLSI, ISO, and EPA protocols with calibration traceability and documentation

Calibration Drift is a Universal Threat to Accuracy

This section illustrated how calibration drift, whether mechanical, electrical, chemical, or environmental in origin, compromises data integrity, operational safety, regulatory compliance, and financial accountability across critical industries. Though drift is often gradual and initially imperceptible, its cumulative effects have led to serious consequences, ranging from patient overexposure and environmental misreporting to economic losses in the billions.

In high-temperature industries like aerospace and power generation, thermocouple drift leads to undetected over-temp conditions due to oxidation and metallurgical instability. In the offshore oil sector, the BP Deepwater Horizon blowout was exacerbated by pressure transducer drift, which masked critical pressure anomalies due to a lack of *in situ* recalibration. In environmental labs, pH electrode miscalibration, often due to expired buffers or poor electrode maintenance, caused incorrect wastewater certifications.

In the healthcare sector, CT scanner dosimetry drift exposed patients to unintended radiation doses due to sensor drift and software update inconsistencies. Meanwhile, at the heart of global metrology, the drift of the International Prototype Kilogram (IPK) relative to its replicas highlighted the limits of artifact-based standards, ultimately leading to the redefinition of the kilogram based on Planck's constant.

In analytical and production environments, instruments showed parallel vulnerabilities: GC-FID systems misreported hydrocarbon concentrations due to column degradation and flame stoichiometry shifts; OES analyzers misclassified alloy grades due to lamp aging and matrix effects; and industrial load cells drifted from thermal, mechanical, and electrical stress, resulting in dosage and fill errors. Spectrophotometers used in paint, plastics, and textiles suffered baseline shifts from the light source and detector degradation, causing false color rejections or approvals.

In custody transfer applications, flow meters drifted due to wear and fouling, costing companies millions in under- or over-measured volumes. In water and biomedical monitoring, ion-selective electrodes (ISEs) misread sodium and chloride concentrations due to membrane degradation and ion interference, jeopardizing both dialysis treatment and environmental compliance.

Despite the diversity of systems and sectors, a consistent pattern emerges: overreliance on initial calibration, long intervals between verification, absence of self-diagnostics, and insufficient environmental compensation. In response, industries adopted corrective measures such as traceable multi-point recalibration routines, redundant measurement systems, real-time diagnostics, and rigorous standards (e.g., ISO/IEC 17025, API MPMS, ASTM, and CLSI).

The overarching lesson is clear: calibration drift is a universal threat to accuracy, but when systematically recognized and proactively managed, it becomes a catalyst for innovation in measurement science, risk control, and instrumentation design.

References

- EPA Method 9211, *Determination of Inorganic Anions by Ion Selective Electrode*, 2004.
- Clinical and Laboratory Standards Institute (CLSI), *C29-A3: Calibration and Quality Control of Sodium and Potassium Ion-Selective Electrode Measurements*, 2018.

- ISO, *Medical laboratories – Requirements for quality and competence (ISO 15189:2022)*. International Organization for Standardization, 2022. The British Standard version (BS EN ISO 15189:2022) has ISBN 978-0-539-02990-1. https://www.en-standard.eu.
- Ives, D.J.G. and Janz, G.J (Eds.), *Reference Electrodes: Theory and Practice*, Academic Press, 1961. ISBN-10: 012376856X, ISBN-13: 978-0123768568.
- Metrohm, A.G., *Best Practices for Ion Selective Electrode Use and Calibration*, 2020.

5

New and Notorious Failures: Analytical Failure and Calibration Risk

Abstract
Seven case studies illustrate failures in analytical integrity across diverse fields. The Theranos scandal weaponized calibration, deceiving patients and regulators. Boeing 737 MAX crashes highlighted reliance on single sensors without cross-validation. Volkswagen's Dieselgate exposed emissions calibration fraud. COVID-19 PCR testing showed misinterpretation and lack of standardization. AI bias in predictive systems demonstrated algorithmic miscalibration. Semiconductor manufacturing faced yield loss from calibration drift and metrology bias. Medical device recalls revealed performance degradation from sensor miscalibration. These cases underscore the critical need for robust calibration practices, transparency, and vigilance to prevent such failures.

Keywords: Theranos scandal, Boeing 737 MAX, Volkswagen Dieselgate, PCR testing, AI bias, semiconductor manufacturing, medical device recalls

The following examples are the consequence of accumulated assumptions, drift in calibration practices, blind trust in systems, and, at times, willful manipulation of measurement itself. This section presents a curated series of seven case studies, each selected for its unique and instructive demonstration of how analytical integrity can break down across fields as diverse as medicine, aviation, artificial intelligence, public health, and advanced manufacturing.

What connects these failures is not their technological content, but their shared vulnerability to analytical erosion, whether through pseudocalibration, misaligned software logic, data bias, or institutional oversight collapse. They expose how trust in models, machines, or measurements can become dangerously misplaced when systems are calibrated for appearance rather than truth and when validation becomes a ritual rather than a safeguard.

- The Theranos scandal reveals how calibration was not merely neglected but weaponized, with fraudulent adjustments

Michael D. Holloway. Accuracy and Calibration Failures in Science and Engineering, (79–104)

designed to mimic accuracy while concealing systemic inaccuracy, resulting in patient harm, regulatory deception, and the erosion of scientific ethics.

- The Boeing 737 MAX disasters demonstrate how reliance on a single sensor without cross-validation led to software-driven crashes. This case underscores the importance of system-level calibration, disagreement logic, and accountability in embedded safety systems.
- Volkswagen's Dieselgate presents a case of deliberate emissions calibration fraud, where software was engineered to perform differently under test conditions than in real-world operation, exposing how engineering can be co-opted for deception under market and regulatory pressure.
- PCR testing during the COVID-19 pandemic illustrates a failure not of instrumentation, but of clinical interpretation and standardization. Over-reliance on high-cycle threshold positivity, absent context or calibration standards, produced inflated case counts and undermined public trust.
- AI bias in predictive systems highlights a new domain of failure: algorithmic calibration, where models appear accurate in aggregate but misfire on subgroups due to biased training data or poorly chosen proxies. These silent miscalibrations embed inequality under the guise of data-driven decision-making.
- Semiconductor manufacturing at the atomic scale exemplifies how calibration drift and metrology bias, even at the angstrom level, can result in massive yield loss, latent chip defects, or reliability risks in mission-critical electronics, reminding us that measurement precision must evolve alongside material and process innovation.
- Medical device recalls involving infusion pumps and ventilators reveal how sensor miscalibration, firmware errors, and alarm system failures can quietly degrade clinical performance, leading to fatal consequences even in devices certified for life support.

Together, these cases form a multidisciplinary map of analytical risk. They reveal how systems fail when calibration is neglected, misunderstood, or exploited. Each study serves not only as a retrospective analysis but as a warning and a call to elevate standards, transparency, and vigilance in how we measure, monitor, and make decisions, whether with instruments, algorithms, or engineered systems.

The following are the details of these examples with root cause and corrective action.

The Theranos Scandal – Weaponized Pseudocalibration in Medical Diagnostics

The Theranos scandal stands as one of the most notorious examples of scientific misconduct in the 21st century, and it exemplifies a particularly insidious failure of analysis: weaponized pseudocalibration. Founded in 2003 by Stanford dropout Elizabeth Holmes, Theranos claimed to revolutionize blood diagnostics with a proprietary device known as the *Edison*. This device, according to Holmes, could perform over 240 tests from a single fingerprick of blood. At its peak valuation of $9 billion, Theranos had attracted a star-studded board of directors and partnerships with major companies such as Walgreens. Yet, the technology never worked as claimed. Instead, Theranos relied on deliberate deception, manipulating calibration procedures and engineering fake analytical precision. The scandal eventually led to criminal convictions, lab closures, and the collapse of the company by 2018 (Carreyrou, 2018; U.S. v. Holmes, 2022).

Pseudocalibration, in general, refers to the act of adjusting an instrument or data processing method to appear accurate without referencing a validated standard. When this approach is employed knowingly either to secure investment, regulatory approval, or user trust, it becomes *weaponized*. At Theranos, pseudocalibration was used not to improve performance but to create the illusion of accuracy. The Edison machines were "trained" to produce expected results rather than derive true values from novel samples. Analysts inside the company routinely adjusted machine outputs to match those from established third-party instruments, such as the Siemens ADVIA 1800, while hiding discrepancies or deleting outliers (FDA, 2015; CMS, 2016).

The roots of this analytical failure were multifold. First, the company lacked validated calibration protocols altogether. Devices were operated with minimal documentation, and instead of calibrating Edison systems against certified reference materials, engineers adjusted outputs to hit anticipated ranges, what one might call "target calibration." No traceability to CLIA or FDA-approved standards existed, violating both scientific and legal norms (CMS, 2016; FDA, 2015).

Second, Theranos frequently employed data cherry-picking. When Edison results failed to align with known laboratory values, the data were omitted from validation studies. Only favorable results were reported, giving the appearance of statistical correlation where none existed. This process was further bolstered by using data generated on traditional third-party instruments while misrepresenting them as produced by the Edison device. This mixture of fraudulent reporting and selective omission produced an apparent, but false, validation narrative (Carreyrou, 2018; SEC, 2018).

A third and critical failure was organizational in nature. Internal whistleblowers like Tyler Shultz and Erika Cheung reported persistent failures in calibration and validation processes, but their concerns were met with intimidation and suppression. Instead of addressing technical concerns, company leadership, including Holmes and COO Sunny Balwani, created a toxic environment where dissent was punished. Employees were required to sign strict NDAs, limiting the possibility of external audit or peer review (Congressional Testimony, 2019).

Further exacerbating the situation was a board of directors and investor base largely composed of individuals lacking medical or scientific expertise. With few, if any, critical questions asked about calibration validation, method uncertainty, or standard reference traceability, the company's narrative went unchallenged. Investors, dazzled by the promise of disruptive innovation, ignored the warning signs of analytical dysfunction (SEC, 2018; Gibbons & Amrhein, 2019).

The mechanisms of pseudocalibration at Theranos can be summarized as follows:

Tactic	Description	Effect
Target Calibration	Devices were "calibrated" to hit *known results*, not *true values*	Created illusion of accuracy
Result Padding	Averaging of accurate third-party data into system "training sets"	Misled on system capability
Mock Validation	Results plotted against legacy lab values, with outliers excluded	Faked correlation and performance
False Claims	Asserted CLIA-equivalent performance without full validation	Violated scientific ethics and law

Root Cause

These methods were not mere mistakes but deliberate strategies that concealed systemic flaws in measurement and analysis. As a result, patients received dangerously inaccurate test results. Some were falsely informed that they had HIV; others were given erroneous pregnancy hormone levels or incorrect blood coagulation indicators. The consequences were real and harmful: lives were endangered, trust in diagnostics was shaken, and regulatory bodies were left scrambling to plug oversight gaps (FDA, 2015; Carreyrou, 2018).

Corrective Action

In the aftermath, multiple corrective actions were proposed and implemented. First, both the FDA and the Centers for Medicare and Medicaid Services (CMS) began tightening oversight of laboratory-developed tests (LDTs). The VALID Act was introduced to bring novel diagnostic platforms under greater pre-market scrutiny, ensuring that devices undergo rigorous validation before clinical deployment. These reforms marked a step forward in reducing the risk of future pseudocalibration scandals (VALID Act Draft, 2020).

Second, regulatory guidance was updated to demand traceability to reference standards issued by organizations such as the National Institute of Standards and Technology (NIST) and the College of American Pathologists (CAP) or compliance with ISO 15189. Instruments and methods now require documentation of calibration intervals, standard operating procedures, and raw calibration data (ISO 15189, 2012).

Third, whistleblower protections within laboratories were expanded. Organizations were urged to implement internal audit systems that tracked calibration changes, logged analytical deviations, and supported employees who raised ethical concerns. Educational initiatives now emphasize ethics in metrology, calibration traceability, and the statistical validity of method comparisons (Institute of Medicine, 2015).

Finally, the Theranos scandal highlighted the importance of competence in oversight. Boards and venture capitalists must include individuals with technical literacy in the relevant fields, particularly in medicine, analytical chemistry, and engineering. Without knowledgeable gatekeepers, future cases of weaponized pseudocalibration may again slip through the cracks.

This case reinforces a central thesis of *When Analysis Gets It Wrong*: calibration is not a clerical task; it is the very skeleton of analytical integrity.

Whether in medicine, aerospace, or environmental testing, any system can be manipulated to "look accurate." The challenge for scientists, engineers, and regulators is to remain skeptical of perfection, vigilant for deception, and uncompromising in the demand for traceability and transparency.

References

- Carreyrou, J., *Bad Blood: Secrets and Lies in a Silicon Valley Startup*, Knopf, 2018.
- U.S. v. Elizabeth Holmes, *et al.*, Case No. 5:18-cr-00258-EJD (N.D. Cal. 2022).
- U.S. Food and Drug Administration (FDA), FDA Form 483 Inspectional Observations, 2015.
- Centers for Medicare and Medicaid Services (CMS), Sanctions and Inspection Report: Theranos Laboratory, 2016.
- U.S. Securities and Exchange Commission (SEC), SEC Charges Theranos, CEO Elizabeth Holmes and Former President Ramesh 'Sunny' Balwani With Massive Fraud, 2018.
- Cheung, E., Congressional Testimony on Scientific Integrity in Labs, 2019.
- Gibbons, R.D. and Amrhein, V., Scientific Misconduct and the Reproducibility Crisis. *Lancet.*, 2019.
- ISO 15189:2012, *Medical Laboratories – Requirements for Quality and Competence.*
- Institute of Medicine, *Improving Diagnosis in Health Care*, 2015.
- VALID Act (Verifying Accurate Leading-edge IVCT Development), Draft legislation introduced in U.S. Congress, 2020.

Boeing 737 MAX – Sensor Disagreement and Software Oversight

The Boeing 737 MAX crisis is a textbook example of how failure to analyze sensor data integrity, software logic, and systemic accountability can lead to catastrophic outcomes. In 2018 and 2019, two brand-new Boeing 737 MAX aircraft, Lion Air Flight 610 and Ethiopian Airlines Flight 302, crashed within five months of each other, killing 346 people. At the center of the disaster was the Maneuvering Characteristics Augmentation System (MCAS), a flight control software intended to compensate for aerodynamic changes caused by the 737 MAX's larger, repositioned engines. However, MCAS relied on a single Angle of Attack (AoA) sensor, and when that sensor provided faulty data, the system activated erroneously, repeatedly pushing the aircraft's nose downward despite pilot efforts to override it (Final Reports: KNKT 2019; EAIB 2020).

What makes this failure notable in the context of analytical breakdowns is not merely the flawed design but the profound misinterpretation and misuse of sensor data. The AoA sensors, which measure the angle between the oncoming air and the wing chord line, are critical to preventing aerodynamic stalls. However, in both fatal incidents, MCAS was triggered by a single faulty AoA reading without cross-verification from the second, redundant sensor available on the aircraft. Boeing's engineers had disabled dual-sensor validation for MCAS to reduce costs and pilot retraining burdens, even though such disagreement checks are standard in safety-critical systems (House Committee Report, 2020).

This lack of cross-checking exemplifies a *sensor disagreement oversight* where disagreement between two or more independent signals is not recognized, logged, or properly acted upon. Instead of recognizing that one sensor was malfunctioning and defaulting to a degraded but stable control mode, the MCAS logic accepted the errant input as truth and forced nose-down trim commands. The underlying assumption that the sensor would always be accurate proved both untested and unsafe.

Additionally, the analytical and testing pipeline failed to detect how this flawed logic would behave under fault conditions. Boeing engineers did not test MCAS activation under scenarios involving a failed AoA sensor, nor did they adequately simulate pilot response times under stress. Internal emails later revealed that some Boeing employees expressed doubts about MCAS safety as early as 2015, but these concerns were neither escalated nor addressed. Worse, Boeing withheld critical information about MCAS from both the FAA and airline customers, including the extent of its authority to move the horizontal stabilizer and the fact that it could activate repeatedly without pilot input (DoJ 2021; NTSB 2019).

The FAA, in turn, delegated significant portions of MAX certification back to Boeing through its Organization Designation Authorization (ODA) program, which allowed Boeing engineers to effectively self-certify critical systems. This created a dangerous conflict of interest and weakened independent oversight, turning safety certification into a bureaucratic formality rather than a rigorous technical review.

The following table summarizes key failure mechanisms:

Failure mode	Description	Consequence
Sensor disagreement ignored	MCAS relied solely on one AoA sensor without validating against the second	Erroneous MCAS activation during normal flight
Unbounded software control	MCAS could repeatedly trim the nose down without pilot override after a fault	Aircraft entered an unrecoverable dive
Inadequate fault simulation	Failure modes were not fully tested under real-world pilot conditions	Delayed recognition and response during emergencies
Omission from Manuals	MCAS was not described in flight crew documentation	Pilots were unaware of the system's existence
Certification gaps	FAA delegated safety review back to Boeing (ODA conflict)	Critical oversight function compromised

Root Cause

The root cause of the 737 MAX disaster is not solely a design flaw. It was a multi-layered systemic failure, beginning with an incomplete risk assessment, compounded by overconfidence in automation, and exacerbated by regulatory abdication. The MCAS system was created as a workaround to preserve the MAX's operational similarity to previous 737 models, avoiding costly level D simulator training. This corporate objective ultimately eclipsed engineering ethics and data integrity standards (Joint Authorities Technical Review, 2019).

Corrective Action

The corrective actions undertaken after the two crashes involved a full redesign of MCAS. Boeing implemented dual-sensor comparison logic, ensuring MCAS would only activate if both AoA sensors were in close agreement. Additional flight deck alerts were mandated, giving pilots real-time information about sensor disagreements. MCAS's ability to repeat commands without pilot input was also removed. More broadly, the FAA reassumed direct oversight of critical systems on the 737 MAX and other aircraft, significantly revising the ODA structure (FAA, 2020).

Furthermore, international aviation authorities, including EASA and Transport Canada, conducted independent reviews of the 737 MAX's airworthiness before recertification, reflecting a loss of trust in the previous U.S.-centric certification process. Pilot training programs were also updated to include MCAS failure scenarios, sensor disagreement handling, and manual stabilizer trim recovery (EASA, 2020; FAA, 2021).

This case teaches crucial lessons in analytical integrity. First, all sensor-dependent systems must include disagreement logic, not as an add-on, but as a baseline safety feature. Second, software that has authority over flight-critical systems must be constrained with bounded behavior and pilot override capability. Third, analytical validation must extend beyond nominal cases and include realistic fault scenarios and human factor modeling. Finally, the importance of independent oversight, particularly in life-critical systems, cannot be overstated.

The 737 MAX disaster is a story not only of technical failure but also of *institutional blindness*, a stark reminder of what happens when analysis is distorted by profit motive, groupthink, and overreliance on unverified assumptions.

References

- KNKT, *Final Report: Lion Air Flight 610*, 2019, Indonesian National Transportation Safety Committee.
- EAIB, *Final Report: Ethiopian Airlines Flight 302*, 2020, Ethiopian Aircraft Accident Investigation Bureau.
- House Committee on Transportation and Infrastructure, *Final Report: The Design, Development, and Certification of the Boeing 737 MAX*, 2020.
- NTSB, *Safety Recommendation Report: Assumptions Used in Safety Assessments for the Boeing 737 MAX Flight Control System*, 2019.
- Joint Authorities Technical Review (JATR), *Boeing 737 MAX Flight Control System Review*, 2019.
- Federal Aviation Administration (FAA), *Summary of FAA Review of the Boeing 737 MAX*, 2020.
- European Union Aviation Safety Agency (EASA), *Return to Service Requirements for the Boeing 737 MAX*, 2020.
- U.S. Department of Justice (DoJ), *Deferred Prosecution Agreement: United States v. The Boeing Company*, 2021.

Volkswagen Dieselgate – Emissions Calibration and Ethical Collapse

The Volkswagen "Dieselgate" scandal represents one of the most flagrant examples of deliberate analytical manipulation through fraudulent calibration and software deceit. Beginning in 2009 and lasting through 2015, Volkswagen (VW) equipped more than 11 million diesel vehicles worldwide, primarily fitted with 2.0-liter TDI engines, with software designed to cheat emissions tests. These "defeat devices," embedded in the engine control unit (ECU), were engineered to detect laboratory test conditions and activate full emissions controls only during those periods. Under real-world driving conditions, nitrogen oxide (NO_x) emissions were up to 40 times higher than legal U.S. limits (EPA, 2015; ICCT, 2014).

At the core of this deception was a dual calibration strategy masked within software logic. The ECU operated in two distinct modes: "dyno mode," which activated full emissions control systems (exhaust gas recirculation, urea injection, etc.) during lab testing, and "road mode," which disabled or drastically reduced those systems to improve fuel economy and performance. This selective calibration violated not only engineering ethics but also national and international environmental laws. More insidiously, it constituted a form of weaponized calibration where calibration settings are manipulated not to improve accuracy but to deliberately mislead regulators, consumers, and the public (House Energy & Commerce Committee, 2016).

The analytical failure here was not a miscalculation or untested assumption; it was intentional obfuscation. Engineers and software developers at VW programmed the defeat devices to identify test conditions based on variables such as steering wheel movement, barometric pressure, vehicle speed, and duration of engine operation. If all conditions suggested a laboratory test (e.g., a fixed steering wheel and specific load profile), the ECU would engage "test mode." In all other cases, the emissions systems were relaxed or bypassed entirely. This false calibration created the illusion of regulatory compliance without actually reducing harmful pollutants during everyday driving (EPA, 2015; DOJ, 2017).

The organizational and ethical collapse extended beyond engineering. Internal documents, uncovered through investigations, revealed that VW executives were aware of the non-compliant behavior as early as 2006. Yet, rather than confront the challenges of meeting emissions standards honestly, the company chose to exploit regulatory blind spots and delay disclosure. Engineers who raised concerns were ignored or pressured into silence. In many ways, the calibration was not just a technical setting; it was a corporate policy disguised as software (Der Spiegel, 2017; New York Times, 2015).

The following table summarizes the failure mechanisms involved:

Failure mode	Description	Consequence
Defeat device logic	Software identified test scenarios and activated emissions control only then	Passed lab tests while violating environmental law on the road
Selective calibration	Dual operating modes: one for regulators, one for the road	Misled regulators and consumers
Suppressed internal warnings	Engineers' objections ignored; whistleblowers silenced or dismissed	Continued fraud and expansion to other models
Ethical breach of trust	Corporate decision to hide non-compliance using code	Criminal liability and brand damage
Regulatory oversight gaps	Reliance on lab-only certification without real-world validation	Created a loophole for software-based fraud

Root Cause

The root cause of Dieselgate was a convergence of engineering shortcuts, corporate hubris, and regulatory myopia. VW's leadership sought to dominate the U.S. market by promoting "clean diesel" as an alternative to hybrid vehicles without committing to the technical investments necessary to meet emissions standards. Rather than implement larger urea tanks, dual-injection strategies, or lower engine compression ratios, VW chose to rely on deceptive software to appear compliant while maximizing vehicle performance and mileage.

The consequences were historic. The U.S. Environmental Protection Agency (EPA), in cooperation with the California Air Resources Board (CARB), initiated an investigation in 2015 after third-party research by the International Council on Clean Transportation (ICCT) revealed discrepancies between lab and road emissions. This led to one of the largest recalls in automotive history, criminal charges against VW executives, and over $30 billion in fines, buybacks, and legal settlements. Multiple executives were arrested, including former CEO Martin Winterkorn (DOJ, 2020).

Corrective Action

Corrective actions following the scandal were sweeping. First, **emissions testing protocols were overhauled globally**. The European Union introduced the **Worldwide Harmonized Light Vehicles Test Procedure (WLTP)**, replacing outdated lab-only standards with **real-driving emissions (RDE)** measurements that require vehicles to be tested under real-world conditions. In the U.S., the EPA adopted new software audit requirements and began installing portable emissions measurement systems (PEMS) to capture on-road NO_x and particulate output (EPA, 2016).

Second, **corporate compliance programs were restructured**. VW created an independent compliance and risk office and agreed to three years of federal oversight under the terms of a **Deferred Prosecution Agreement**. The company also committed to a corporate integrity monitor to enforce ethical standards and transparency in engineering decisions (DOJ, 2017).

Finally, engineers, analysts, and regulators globally have learned the hard lesson that **calibration can be corrupted when business interests override ethical standards**. Any analytical system, particularly one involving regulation, public health, or environmental stewardship, must be subject to **transparent audit, robust real-world validation, and non-negotiable traceability** to physical outcomes. Calibration must reflect objective performance, not strategic deception.

The Dieselgate scandal exemplifies a new form of failure **calibration fraud as a strategic weapon** and reveals that the line between analytical rigor and corporate manipulation can be as thin as a few lines of code.

References

- U.S. Environmental Protection Agency (EPA), *Notice of Violation to Volkswagen AG, Audi AG, and Volkswagen Group of America*, 2015.
- International Council on Clean Transportation (ICCT), *Real-World Exhaust Emissions from Modern Diesel Cars*, 2014.
- U.S. Department of Justice (DOJ), *Plea Agreement and Statement of Facts: United States v. Volkswagen AG*, 2017.

- House Committee on Energy and Commerce, *VW Emissions Investigation: Preliminary Staff Report*, 2016.
- Der Spiegel, The Code of Deception: How Software Engineers at VW Manipulated Diesel Engines, 2017.
- The New York Times, Volkswagen Says 11 Million Cars Worldwide Are Affected in Diesel Deception, 2015.
- EPA, *Final Rule: Light-Duty Vehicle Testing Procedures and Emission Requirements*, 2016.
- European Union, *Worldwide Harmonized Light Vehicles Test Procedure (WLTP) and Real Driving Emissions (RDE) Requirements*, 2018.
- DOJ, *Volkswagen AG Completion of Deferred Prosecution Agreement*, 2020.

PCR Testing During COVID-19 – False Positives and Drift in Clinical Practice

Polymerase chain reaction (PCR) testing became the cornerstone of global diagnostic strategies during the COVID-19 pandemic. Initially praised for its high sensitivity and specificity in detecting SARS-CoV-2 viral RNA, PCR testing rapidly scaled from research settings to mass clinical deployment. Yet, as the pandemic evolved, PCR testing faced growing scrutiny over false positives, cycle threshold (Ct) drift, and misuse in public health decision-making. What began as a gold-standard molecular method eventually exposed systemic weaknesses in analytical interpretation, standardization, and clinical calibration.

PCR tests work by amplifying fragments of viral genetic material through repeated thermal cycles. Each cycle doubles the amount of nucleic acid present. The cycle threshold (Ct) value refers to the number of cycles required for a detectable signal to emerge. Lower Ct values correlate with higher viral loads; higher Ct values indicate either very low viral presence or, potentially, environmental contamination or artifact (Mina *et al.*, 2020). In theory, tests with a Ct value above a certain threshold, often 35 to 40, should be considered inconclusive or invalid, especially if viral replication competency is questionable.

However, as laboratories and hospitals scrambled to increase testing capacity, standardized Ct thresholds were inconsistently applied across manufacturers, platforms, and regions. Some labs considered any detectable signal as "positive," regardless of Ct value, leading to a rise in clinically irrelevant or false-positive results. In certain jurisdictions, individuals with Ct values of 38–40 were labeled positive and subjected to quarantine even though multiple studies suggested such individuals were non-infectious (Bullard *et al.*, 2020; Tom & Mina, 2020).

The issue was compounded by a drift in clinical interpretation and regulatory guidance. Initial emergency authorizations from agencies like the U.S. FDA permitted flexible assay development under the Emergency Use Authorization (EUA) pathway. This resulted in a proliferation of assays with non-uniform calibration, unverified performance across Ct ranges, and limited real-world validation (FDA, 2021). Laboratories often lacked access to standard reference materials, resulting in divergent test sensitivities and inconsistencies in viral load reporting. Without centralized calibration to World Health Organization (WHO) standards or NIST-traceable controls, cross-comparability across test kits eroded (WHO, 2020).

The following table outlines critical failure mechanisms observed during this period:

Failure mode	Description	Consequence
Ct threshold drift	Varying cutoffs across labs and platforms without consensus	Increased false positives, inconsistent infectivity determination
Lack of Calibration Standards	Absence of common viral RNA reference materials or quantification controls	Inconsistent result interpretation across time and location
Over-interpretation of low viral loads	Positive calls made on Ct >38 without clinical correlation	Unnecessary isolation and epidemiological distortion
Decoupling from clinical context	Testing decisions divorced from symptoms or transmission likelihood	Erosion of diagnostic utility and public trust
Regulatory laxity under EUA	Assay variation and limited oversight during early scaling	Method drift and confusion among practitioners

Root Cause

The **root cause** of this analytical failure lies in the intersection of **method sensitivity exceeding clinical relevance** and the **lack of harmonized interpretation guidelines**. PCR is exquisitely sensitive and capable of detecting trace RNA sometimes long after the virus is no longer viable. However, in the absence of Ct-based cutoffs grounded in infectiousness, the test was often misapplied as a binary diagnostic tool rather than part of a broader clinical picture. The use of extremely high Ct values without confirmatory clinical assessment created false positive cases, drove unnecessary quarantines, and potentially inflated case counts during the early waves of the pandemic (Tom & Mina, 2020).

Another layer of failure was **epistemic drift**, the subtle shift in how PCR was understood by clinicians, policymakers, and the public. Originally designed as a molecular tool for confirmation in symptomatic patients, PCR was rapidly repurposed into a universal screening tool, sometimes applied indiscriminately. Over time, the clinical nuance around Ct values and viral shedding was lost, and PCR results were interpreted without context, transforming a precise tool into a blunt policy instrument.

Corrective Action

Corrective actions included the eventual adoption of WHO guidance in late 2020, which emphasized that "a positive PCR result does not necessarily indicate infectiousness" and encouraged laboratories to report Ct values alongside results when available (WHO, 2021). Several European nations revised public health recommendations, advising that Ct values over 35 be treated with caution and potentially retested in the presence of clinical symptoms. The U.S. CDC began promoting symptom-based return-to-work policies, shifting reliance away from PCR test clearance alone (CDC, 2021).

Further reforms proposed include mandatory calibration of molecular assays to international reference standards, improved assay transparency from manufacturers regarding Ct behavior, and retraining of clinicians in molecular result interpretation. Future pandemic preparedness efforts must include frameworks for analytical standardization, robust data infrastructure for real-time calibration monitoring, and fail-safe designs that flag questionable results (e.g., very high Ct values) for reflex testing or clinician review.

This case underscores the principle that amplification without interpretation is dangerous. A powerful analytical method, when stripped of context and calibration, can undermine the very goals it was deployed to support. In the case of PCR during COVID-19, high sensitivity became a double-edged sword, identifying cases long after clinical resolution, misinforming public health responses, and contributing to societal fatigue over restrictions.

References

- Bullard, J., *et al.*, Predicting infectious SARS-CoV-2 from diagnostic samples. *Clin. Infect. Dis.*, 71, 10, 2663–2666, 2020.
- Mina, M.J., Parker, R., Larremore, D.B., Rethinking Covid-19 Test Sensitivity A Strategy for Containment. *N. Engl. J. Med.*, 383, 22, e120, 2020.
- Tom, M.R. and Mina, M.J., To Interpret the SARS-CoV-2 Test, Consider the Cycle Threshold Value. *Clin. Infect. Dis.*, 71, 16, 2252–2254, 2020.
- U.S. Food and Drug Administration (FDA), *FAQs on Testing for SARS-CoV-2*, 2021.
- World Health Organization (WHO), *Diagnostic Testing for SARS-CoV-2: Interim Guidance*, 2020.
- WHO, *WHO Information Notice for Users 2021/05: Interpretation of PCR results for diagnosis of SARS-CoV-2*, 2021.
- Centers for Disease Control and Prevention (CDC), *Interim Guidance on Duration of Isolation and Precautions for Adults with COVID-19*, 2021.

AI Bias and Algorithmic Calibration: When Models Learn Wrong

Artificial intelligence (AI) systems now guide decisions in areas as sensitive as credit approvals, medical diagnostics, job screening, and criminal sentencing. Central to these systems is algorithmic calibration, the process by which models are tuned to ensure predicted probabilities align with real-world outcomes. Yet, calibration itself can be corrupted. When training data reflect historical inequalities, or when model developers overlook subgroup performance, AI systems can learn patterns that replicate or even amplify societal bias. The result is a new frontier of analytical failure: models that appear precise but are wrong in morally and socially dangerous ways.

One of the most widely cited examples of algorithmic bias occurred in the criminal justice domain. In 2016, an investigative report by *ProPublica* revealed that COMPAS, a proprietary risk assessment tool used to estimate recidivism likelihood, systematically overestimated the risk posed by Black defendants and underestimated the risk for white defendants (Angwin *et al.*, 2016). While COMPAS was technically calibrated, meaning its scores aligned with overall outcomes, it was poorly calibrated across demographic subgroups. A score of "high risk" meant different things for different racial groups. This discrepancy was invisible in global accuracy statistics but had direct consequences for bail, parole, and sentencing decisions.

This case illustrates a form of algorithmic pseudocalibration: models are validated against average outcomes, but hidden biases in input features, label selection, or optimization criteria distort those results. In the COMPAS case, the algorithm used historical arrest records and convictions to predict recidivism, failing to account for the systemic over-policing of certain communities. The model, in effect, learned societal bias and framed it as an objective risk (Binns, 2018; Barocas *et al.*, 2019).

A similar issue emerged in healthcare. In 2019, a major study published in *Science* found that a commercial healthcare risk prediction algorithm used to allocate "extra care" to high-risk patients underestimated the health needs of Black patients by more than 40% (Obermeyer *et al.*, 2019). The root problem lies in the proxy variable chosen for model training: past healthcare costs. Because Black patients historically receive fewer medical services (due to access disparities, mistrust, or discrimination), the algorithm wrongly concluded they were healthier and needed less attention. The model was statistically calibrated to past spending, not to actual health outcomes, producing racially skewed predictions under the guise of precision.

The following table summarizes critical failure mechanisms observed in these cases:

Failure mode	**Description**	**Consequence**
Subgroup miscalibration	The model appears accurate globally but is inaccurate for protected groups	Discriminatory treatment under the color of objectivity
Biased training data	Historical inequality encoded into feature distributions and labels	Algorithms replicate societal prejudice
Faulty proxy variables	Outcome proxies (e.g., cost = health, arrests = danger) entrench systemic bias	Inaccurate predictions for marginalized populations
Opaque model deployment	Proprietary algorithms shielded from audit or external review	Hidden errors persist uncorrected
Calibration blindness	Lack of subgroup calibration checks during validation	Models pass quality review despite ethical flaws

Root Cause

The root cause of these failures lies in a narrow interpretation of accuracy and fairness. Developers often prioritize overall performance metrics such as AUC, precision, or Brier score without disaggregating results by race, gender, or other protected categories. Furthermore, commercial incentives favor speed and scale over equity and explainability. In many domains, algorithmic models are deployed without rigorous fairness auditing, ethical oversight, or stakeholder consultation. The result is a false sense of scientific objectivity, wherein biased predictions masquerade as neutral outputs (Barocas & Selbst, 2016).

Corrective Action

Corrective actions for algorithmic miscalibration are now emerging from both academia and policy circles. One of the key recommendations is the routine use of group-wise calibration curves, which compare predicted risk against actual outcomes across demographic subgroups. Techniques like equalized odds, demographic parity, and fairness through unawareness offer frameworks to evaluate and mitigate bias. Additionally, there is a growing demand for model interpretability, especially in high-stakes domains. Tools like LIME and SHAP allow practitioners to see how specific features influence predictions, making hidden biases more visible (Ribeiro *et al.*, 2016; Lundberg & Lee, 2017).

From a policy perspective, several jurisdictions now require algorithmic impact assessments (AIAs) akin to environmental impact statements for models deployed in public services. The EU's AI Act and U.S. proposals such as the Algorithmic Accountability Act aim to enforce transparency, fairness testing, and independent audit rights for automated decision systems (European Commission, 2021; U.S. Congress, 2022).

The lessons from these cases are clear: calibration without conscience is not enough. Just as poorly calibrated instruments can produce false measurements, poorly calibrated algorithms can produce false realities that carry the imprimatur of science but perpetuate discrimination. In a data-driven world, algorithmic calibration must be both technically sound and ethically robust, or it will become the next frontier of analytical failure hiding in plain sight.

References

- Angwin, J., Larson, J., Mattu, S., Kirchner, L., *Machine bias. ProPublica*, 2016, https://www.propublica.org/article/machine-bias-risk-assessments-in-criminal-sentencing.
- Barocas, S., Hardt, M., Narayanan, A., *Fairness and Machine Learning*, 2019, fairmlbook.org.
- Obermeyer, Z., Powers, B., Vogeli, C., Mullainathan, S., Dissecting racial bias in an algorithm used to manage the health of populations. *Science*, 366, 6464, 447–453, 2019.
- Binns, R., Fairness in Machine Learning: Lessons from Political Philosophy, in: *Proceedings of the 2018 Conference on Fairness, Accountability, and Transparency (FAT)*, 2018.
- Ribeiro, M.T., Singh, S., Guestrin, C., Why Should I Trust You? Explaining the Predictions of Any Classifier, in: *Proceedings of the 22nd ACM SIGKDD*, 2016.
- Lundberg, S.M. and Lee, S.-I., A Unified Approach to Interpreting Model Predictions. *Adv. Neural Inf. Process. Syst.*, 2017.
- Barocas, S. and Selbst, A.D., Big Data's Disparate Impact. *Calif. Law Rev.*, 104, 3, 671–732, 2016.
- European Commission, *Proposal for a Regulation Laying Down Harmonised Rules on Artificial Intelligence (AI Act)*, 2021.
- U.S. Congress, *Algorithmic Accountability Act of 2022*, 2022.

Semiconductor Manufacturing – Calibration Failure at the Atomic Scale

Few industries demand greater precision than semiconductor manufacturing. The success of each chip depends on the exact placement and composition of features at scales measured in nanometers and, increasingly, at the level of **individual atoms**. As transistor nodes shrink below 5 nm, the margin for error in **etching, deposition, doping, and lithographic alignment** becomes vanishingly small. At this level, **calibration is not a routine procedure; it is an existential constraint**. Any drift, noise, or miscalibration can cascade into catastrophic yield loss, latent defects, or functional degradation in finished chips.

At the heart of the calibration challenge lies **process control**. In modern fabs, equipment such as atomic layer deposition (ALD) chambers, electron-beam lithography systems, and metrology tools like scanning electron microscopes (SEM) and atomic force microscopes (AFM) must all operate in tightly coordinated harmony. Measurements of film thickness, critical dimensions (CDs), line edge roughness, and dopant concentration must be accurate to within **angstroms (10^{-10} m)**. Any deviation, no matter how small, can result in incorrect layer stacking, misaligned vias, or transistor leakage that escapes detection until the final test or, worse, in-field failure (ITRS, 2015).

A striking example occurred during the 2010s in advanced FinFET fabrication. Engineers at one facility observed unexplained **device instability** and premature aging during post-fab burn-in tests. The culprit was traced to **slight over-doping of the gate region**, which subtly altered the threshold voltage in high-density transistors. Root cause analysis revealed that an **ion implanter had drifted from its intended dosage calibration over a 72-hour period**, a deviation of just 2–3%, undetectable without atomic-scale dose monitoring. Compounding the issue, the process control system had disabled real-time feedback due to a misconfigured sensor validation flag (SEMATECH, 2016).

This case represents a class of failure known as **nano-calibration drift,** in which instrumentation remains "in spec" on paper yet fails to meet real-world production tolerance due to the ultra-sensitive nature of the process. Conventional Six Sigma methodologies, often designed for ±1% tolerances, are inadequate at atomic scales. Furthermore, many in-line metrology systems suffer from **cumulative bias**, where small sensor or stage alignment errors add up over time. If calibration routines are not performed with sub-nanometer traceability to primary standards, **"false normalcy" sets in**, and systems may report success while silently drifting out of spec.

The following table summarizes key calibration-related failure modes in atomic-scale semiconductor processes:

Failure mode	Description	Consequence
Nano-scale calibration drift	Gradual deviation in tool output undetectable by standard tolerance checks	Subtle electrical failure, increased chip variability
Cumulative sensor bias	Slight misalignment or sensor degradation aggregates over many wafers	Systemic yield loss or process excursions
Feedback loop failure	Real-time process correction disabled due to software or signal configuration	Uncorrected deviations in film thickness or doping profile

(*Continued*)

(*Continued*)

Failure mode	Description	Consequence
Pseudo-stability masking	Tool outputs appear stable within spec but deviate from true calibration	Undetected process variation and product-level reliability risk
Metrology blind spots	Missing inline checks on intermediate layers or doping uniformity	Latent failures not caught until final electrical testing

Root Cause

The root causes of calibration failure at this scale include not only technical drift but also limits in measurement resolution, improper sensor validation logic, and incomplete modeling of process interactions. In ultra-thin layers (<5 nm), even quantum mechanical effects like electron tunneling can interfere with traditional metrology assumptions. Moreover, tool vendors often provide calibration systems that rely on secondary references, which may not be traceable to SI-unit primary standards like those from NIST. As a result, fabs may operate under a veil of precision that is more illusory than absolute.

Corrective Action

To address these challenges, the semiconductor industry has initiated several corrective actions. Leading fabs (e.g., TSMC, Intel, and Samsung) now employ multi-layer cross-tool calibration chains, where measurements from ellipsometry, X-ray photoelectron spectroscopy (XPS), and transmission electron microscopy (TEM) are compared across processes and machines. Tools are benchmarked daily against nanoscale calibration wafers with embedded standards traceable to national metrology institutes. Additionally, AI-driven statistical process control (SPC) systems are now used to detect subtle drift patterns before they manifest in die-level variation (ASML, 2020).

An equally important corrective measure is closed-loop real-time feedback. Advanced deposition and etching tools are now equipped with optical emission spectroscopy (OES) and real-time interferometry sensors, which allow the process to adjust in flight. However, these systems require precise calibration themselves, meaning the chain of trust must extend from the measuring instrument all the way to the atomic feature on the wafer.

Finally, the industry is developing next-generation calibration standards, including atomically precise step-height gauges and dopant concentration references verified by neutron activation analysis. Organizations like NIST, VLSI Standards, and SEMI are collaborating to ensure that calibration practices keep pace with device miniaturization and materials innovation.

The lesson is clear: at atomic scales, even the smallest miscalibration becomes consequential. The illusion of control reinforced by unverified feedback loops or incomplete metrology can give rise to systemic failure hidden beneath layers of statistical "compliance." As devices become smaller, more complex, and more interconnected, the integrity of calibration itself becomes the most critical variable in ensuring performance, safety, and yield.

References

- International Technology Roadmap for Semiconductors (ITRS), *Metrology Chapter*, 2015.
- SEMATECH, *Ion Implantation Dose Error Case Study Report*, 2016.
- ASML, *Advanced Lithography and Inline Calibration Technologies*, 2020.
- National Institute of Standards and Technology (NIST), *Nanoscale Metrology Program Overview*, 2021.
- VLSI Standards Inc., *Calibration Wafers and Metrology Artifacts*, 2020.
- SEMI, *Semiconductor Equipment and Materials International: Calibration Guidelines for Advanced Node Manufacturing*, 2021.

Medical Device Recalls – Infusion Pumps and Ventilators

Medical devices are held to some of the highest standards of safety and reliability because their failure can lead directly to patient injury or death. Yet, over the past two decades, thousands of infusion pumps and mechanical ventilators have been recalled for critical issues tied not only to mechanical faults but also to **analytical failures** chief among them **miscalibration, sensor drift**, and flawed software control logic. These cases illuminate the fragility of life-critical devices when calibration, verification, and system-level validation are insufficiently robust or neglected under commercial pressure.

Between 2005 and 2010, infusion pump recalls surged, culminating in a *landmark FDA initiative* called the Infusion Pump Improvement Initiative. During this period, the FDA received over **56,000 reports of adverse events** associated with infusion pumps, including **500 deaths**. Common failure points included incorrect dosage delivery, unrecognized occlusions, and failure of flow sensors. The root cause in many cases was a **drift in flow rate calibration**, often combined with **software errors in dosage calculations or alarm-silencing logic** (FDA, 2010).

One prominent case involved the **Hospira (later Pfizer) Symbiq infusion system**, which was recalled after several incidents where incorrect therapy parameters were transmitted from a networked software interface to the device. The infusion pumps **misinterpreted drug libraries or overrode safety limits**, leading to inappropriate flow rates. At the heart of the issue was a **lack of synchronization between software modules and device calibration settings**, resulting in what amounted to silent failures—**errors that did not trigger alarms** but still caused harm (FDA Recall, 2015).

In ventilators, a particularly alarming recall occurred during the COVID-19 pandemic in 2020–2021, involving multiple manufacturers and models. A case in point: Philips Respironics recalled millions of CPAP, BiPAP, and ventilator devices due to **polyester-based polyurethane sound abatement foam** degrading into the airway path. While initially regarded as a materials issue, the failure was compounded by **insufficient sensor calibration to detect pressure anomalies** associated with internal contamination and **algorithmic failure to trigger alarms during compromised airflow** (Philips Safety Notice, 2021). These events highlighted how **physical component degradation**, if not properly sensed or interpreted, can elude detection in devices meant to support or replace vital human functions.

The table below summarizes common failure modes observed in these device classes:

Failure mode	**Description**	**Consequence**
Calibration drift	Sensors measuring flow rate, pressure, or volume slowly deviate from reference	Incorrect drug dosing or respiratory support
Software override errors	Programming logic overrides safety limits or ignores calibration data	Critical infusion or ventilation parameters misapplied

(*Continued*)

(*Continued*)

Failure mode	**Description**	**Consequence**
Alarm fatigue/ suppression	Alarms not calibrated to true failure conditions or silenced by false positives	Device fails silently, user unaware of malfunction
Networked device sync gaps	Parameters sent from external interfaces not synced with device calibration	Dose miscalculation or safety limit bypass
Degraded material undetected	Sensor systems fail to recognize internal degradation or contamination	Airflow blockages or toxic exposure without alarm activation

The **root causes** of these recalls are traced back to a combination of **design complexity, incomplete failure mode analysis, inadequate calibration maintenance, and insufficient software-hardware integration testing**. Many infusion pumps and ventilators were not built with adequate redundancy in sensor systems or with traceable calibration verification routines that could flag drift or degradation before patient harm occurred. Often, testing focused on nominal conditions, not rare or edge-case combinations where drift, wear, or data mismatches could create cascading failures (ECRI Institute, 2016).

Corrective actions that have since been implemented or proposed include mandatory **traceability of calibration to NIST or ISO standards** and requirements for **real-time calibration checks** for critical parameters such as pressure, flow rate, or dose volume. Device manufacturers are also encouraged to implement **watchdog systems** that monitor sensor agreements, detect abnormal drift patterns, and require **multi-sensor validation** for any decision affecting life-critical flows (AAMI, 2021).

From a regulatory standpoint, the FDA now mandates that device software and firmware undergo **rigorous cybersecurity and communication pathway validation**, especially when external programming interfaces are involved. Device manufacturers must submit **Human Factors Engineering (HFE) reports** to demonstrate that alarm systems, displays, and overrides are intuitive, fail-safe, and resistant to misinterpretation or misuse (FDA Guidance, 2016).

The recurring theme in these failures is not only mechanical fault or software bugs but also a deeper **failure of analytical design assurance**, an assumption that once-calibrated devices will remain accurate without continuous verification and that users will interpret alarms and overrides correctly even under stress. These assumptions fail when devices are used in chaotic, high-stakes environments like ICUs, operating rooms, or pandemic triage centers.

Ultimately, **calibration is not static;** it must be **continually verified, interpreted correctly by software, and understood by end users**. The cost of ignoring this is measured not only in product recalls but also in lives.

References

- U.S. Food and Drug Administration, *White paper: Infusion Pump Improvement Initiative.* U.S. Department of Health & Human Services, 2010, Retrieved from https://www.fda.gov/media/79010/download.
- U.S. Food and Drug Administration, *Safety communication: Cybersecurity vulnerabilities of Hospira Symbiq Infusion System (Class I recall)*, 2015, July 31, Retrieved from https://www.fda.gov/medical-devices/recalls/hospira-symbiq-infusion-system.
- Philips Respironics, *Medical device correction: CPAP and ventilator recall notice*, 2021, June, Retrieved from https://www.philips.com/a-w/about/news/archive/standard/news/press/2021/20210614.
- ECRI Institute, *Top 10 Health Technology Hazards for 2016.* ECRI Institute, 2015, November, Retrieved from https://www.ecri.org/press/Pages/Dirty_Endoscopes_Top_ECRI_Institutes_2016_Technology_Hazards_List.aspx?_cldee=amJ1cmdlckBvdXRwYXRpZW50c3VyZ2VyeS5uZXQ%3d.
- Association for the Advancement of Medical Instrumentation (AAMI), *Technical Information Report on Calibration Verification Practices in Medical Devices*, 2021.
- FDA, *Guidance for Industry and FDA Staff: Applying Human Factors and Usability Engineering to Medical Devices*, 2016.

Summary of New and Notorious Failures

This section examined high-profile, recent failures where calibration or its distortion, absence, or mismanagement played a pivotal role in technological breakdowns, ethical violations, and systemic risk exposure. While spanning diverse domains, each case reflects how calibration whether technical, algorithmic, or procedural is a critical safeguard that, when compromised, transforms complexity into catastrophe.

In the Theranos scandal, the company deliberately fabricated calibration protocols to make its blood testing devices appear functional. By simulating precision through "pseudocalibration," the technology deceived regulators, investors, and patients, weaponizing false data to build trust where none was earned. In aerospace, the Boeing 737 MAX crashes were the tragic result of sensor disagreement and software override logic where one faulty angle-of-attack sensor, unchecked by redundancy or drift detection, triggered MCAS and led to two fatal crashes.

The Volkswagen Dieselgate affair illustrates an ethical collapse in calibration governance: engineers manipulated engine control software to detect emissions tests and alter combustion parameters for cleaner lab performance while emitting illegal levels of NOx in real-world driving. This deliberate miscalibration undermined regulatory trust and caused billions in fines, recalls, and reputational damage.

During the COVID-19 pandemic, the global dependence on PCR testing revealed issues with false positives, over-sensitivity, and procedural drift. Misaligned Ct (cycle threshold) values, inconsistent lab calibration, and lack of standardization distorted case counts and public health decisions. In parallel, AI systems trained on unbalanced datasets began to exhibit biased decision-making a form of algorithmic drift, where models internalized prejudiced patterns due to poorly calibrated training inputs and feedback loops.

In advanced manufacturing, particularly semiconductor fabrication, drift occurs at the atomic scale, where instruments like lithography and etch tools require sub-nanometer calibration to maintain yield. Even minuscule deviations in alignment, temperature, or mask overlay create billions in losses due to defective chips. Finally, in medical device recalls involving infusion pumps and ventilators, calibration neglect resulted in over-infusion, incorrect flow rates, and failure to detect occlusions impacting patient safety and prompting widespread recalls.

Across all cases, the common thread is that calibration once thought of as a routine technical detail is increasingly a strategic, ethical, and safety-critical function. Failures here don't just erode measurement accuracy; they erode trust, expose liability, and reveal deeper fractures in design thinking, regulatory compliance, and ethical oversight.

6

Bad Data Leads to Bad Things Beyond the Lab

Abstract

The Challenger disaster exemplifies catastrophic outcomes from flawed data interpretation and communication failures. Engineers warned of O-ring vulnerabilities in cold weather, but sparse, poorly presented data led to dismissal by NASA leadership. The launch proceeded under schedule pressure, resulting in O-ring failure and the shuttle's destruction. This incident underscores the dangers of incomplete data, inadequate communication, and prioritizing schedules over safety. It serves as a critical lesson in engineering ethics, risk assessment, and the need for transparent dialogue between technical teams and decision-makers across high-stakes industries.

***Keywords*:** Challenger disaster, chernobyl disaster, data integrity, risk communication, engineering ethics, decision-making, system limits

There are many powerful, sobering examples in history where decisions made on flawed or misunderstood data led to catastrophic outcomes. These stories are used across disciplines from engineering and medicine to finance and military planning as cautionary tales about the importance of data integrity, validation, and context. Here are some of the most notable.

Challenger Space Shuttle Disaster (1986)

- Failure Point: Misinterpreted or dismissed temperature-related data.
- Details: Engineers warned that O-ring seals on the solid rocket boosters might fail in cold weather. However, data on O-ring behavior was sparse and not clearly presented to NASA leadership.

Michael D. Holloway. Accuracy and Calibration Failures in Science and Engineering, (105–120)

- Consequence: The O-rings failed at launch due to cold temperatures, leading to the breakup of the shuttle and the death of all 7 crew members.
- Lesson: Incomplete data + pressure to proceed = catastrophic decision. It also shows the danger of poor communication between technical staff and decision-makers.

The Challenger disaster remains one of the most defining tragedies in aerospace history, not only for the loss of seven astronauts but also for its profound implications on engineering ethics, data communication, and risk-based decision-making. On the morning of January 28, 1986, the Challenger shuttle broke apart 73 seconds after launch, a failure traced back to the O-ring seals in the solid rocket boosters (SRBs), small but crucial components that were not designed to perform reliably in the cold conditions present at the launch.

The failure point was known, at least in part, before the incident. Engineers from Morton Thiokol, the contractor responsible for the SRBs, had warned that the O-rings, rubber-like gaskets designed to seal joints, were vulnerable at low temperatures. Specifically, they feared that in cold weather, the O-rings could become too brittle to seal properly, allowing hot combustion gases to escape and compromise the integrity of the booster. However, this concern was not supported by comprehensive, statistically significant data, and what data did exist was inconsistently presented. Graphs lacked clear annotations, and the relationships between temperature and O-ring failure were not effectively modeled or communicated. Moreover, the historical data set on O-ring performance at temperatures below freezing was sparse, making it easier for decision-makers to discount the warnings as inconclusive.

NASA, under intense pressure to maintain a tight launch schedule, was dealing with a backlog of shuttle flights, public scrutiny, and political expectations, including the high-profile nature of Challenger's mission, which included Christa McAuliffe, the first civilian teacher in space. These pressures contributed to a culture of normalization of deviance, where known risks became acceptable because previous launches had not failed despite similar issues. In a now-infamous conference call on the eve of the launch, engineers recommended delaying the flight due to the forecast cold temperatures, but senior managers at Morton Thiokol reversed the engineers' recommendation under pressure, and NASA proceeded with the launch.

The O-rings failed precisely as the engineers feared. The temperature at launch was 36°F (2°C), far lower than any previous launch. The O-rings did not properly seal, and hot gases escaped through the SRB joint, eventually breaching the external fuel tank, and leading to the catastrophic breakup of the shuttle.

The consequences of the Challenger disaster extended beyond the immediate tragedy. It catalyzed a complete overhaul of NASA's safety culture, decision-making structure, and engineering review processes. The Rogers Commission, which investigated the disaster, cited not just the technical failure of the O-rings but also the organizational failure to respond adequately to concerns raised by engineers. Nobel laureate Richard Feynman, a member of the commission, famously demonstrated the O-ring's brittleness in ice-cold water, emphasizing how engineering concerns were downplayed in favor of schedule-driven pressures.

The core lesson of the Challenger disaster is clear and enduring: incomplete or poorly communicated data, combined with institutional pressure to proceed, can lead to catastrophic consequences. It is a sobering reminder of the importance of transparent communication between technical teams and decision-makers, the ethical responsibility to elevate safety over schedules, and the necessity of treating uncertainty and dissenting opinions as signals, not obstacles, to sound decision-making.

This tragedy underscores why measurement integrity, risk communication, and ethical engineering culture must be prioritized not just in spaceflight, but in every high-stakes industry where lives and trust are on the line.

Chernobyl Nuclear Disaster (1986)

- Failure Point: Overriding safety protocols based on faulty assumptions and misunderstood reactor behavior.
- Details: Operators conducted a test without fully understanding the reactor's unstable state. Data indicating dangerous power fluctuations was misinterpreted.
- Consequence: Explosion and meltdown, widespread radioactive contamination, long-term health/environmental consequences.
- Lesson: Data without context is dangerous; training, simulation, and understanding system limits are vital.

The Chernobyl disaster, one of the worst nuclear accidents in history, occurred on April 26, 1986, at the Chernobyl Nuclear Power Plant in Pripyat, Ukraine (then part of the Soviet Union). The explosion and meltdown of Reactor No. 4 released massive quantities of radioactive materials into the atmosphere, resulting in widespread contamination, long-term environmental damage, and a still-growing toll of health-related consequences, particularly cancers and radiation sickness, among workers and residents.

At the core of the disaster was a tragic and preventable failure to understand, respect, and properly respond to data and system behavior. The reactor in question was an RBMK-1000 design, known for certain inherent instabilities at low power and a positive void coefficient, meaning that as steam bubbles (voids) form in the reactor coolant, reactivity increases instead of decreases, potentially leading to runaway reactions.

The failure point was rooted in a poorly designed safety test and compounded by critical gaps in operator training, instrumentation interpretation, and reactor behavior understanding. The test aimed to determine whether the turbine generator could provide enough residual power to keep coolant pumps running during a power loss, just before emergency systems kicked in. However, during test preparations, operators manually overrode multiple safety systems, including the automatic shutdown protocols, and ran the reactor at dangerously low power levels, a condition that made it unstable and highly reactive.

Complicating matters, data from control instruments showed irregular power fluctuations and rising reactivity, but operators, lacking a full understanding of RBMK dynamics and trusting in the perceived safety of their procedures, misinterpreted or dismissed these warning signs. Some warning systems were even disabled, as they were seen as nuisances during the test. The operators assumed they were in control, but the reactor's internal physics told a different story.

When the operators finally initiated an emergency shutdown by inserting control rods in a last-ditch attempt to stabilize the reactor, the rods themselves triggered a dramatic power spike due to their flawed design. The sudden spike led to a steam explosion, followed by a second hydrogen explosion, blowing the 1,000-ton reactor lid off and exposing the core. The result: a meltdown and fire that burned for days, spewing radioactive isotopes across Ukraine, Belarus, Russia, and much of Europe.

The consequences were devastating. Immediate casualties included plant workers and first responders, many of whom died from acute radiation syndrome within weeks. Long-term impacts include thousands of cases of thyroid cancer, extensive evacuations, an exclusion zone still uninhabitable today, and incalculable economic and ecological costs. The event also eroded public trust in nuclear energy and exposed severe weaknesses in Soviet safety culture, engineering oversight, and communication transparency.

The lesson of Chernobyl is multi-layered and far-reaching: data without context is dangerous. The operators were looking at real data on reactor power levels, coolant flow, and control rod positions but lacked the training, system knowledge, and interpretive framework to understand what that data meant under unstable conditions. In fact, it was not just a technical failure; it was a cognitive and cultural failure. The disaster highlighted the dangers of overriding automated protections, the risk of complacency in complex systems, and the catastrophic cost of acting on assumptions rather than understanding.

To prevent such events in any industry, whether in nuclear energy, aerospace, pharmaceuticals, or manufacturing, organizations must foster deep system literacy, rigorous scenario-based training, and a culture of questioning. Instruments can only measure what they're designed to sense; it is up to the humans behind the controls to interpret, challenge, and respond wisely. At Chernobyl, the absence of that awareness turned data into a deadly illusion and a reactor into a bomb.

Three Mile Island Nuclear Incident (1979)

- Failure Point: Misleading control panel data.
- Details: A valve stuck open, but the control room display falsely indicated it was closed. Operators misdiagnosed the issue and worsened the problem.
- Consequence: Partial core meltdown; significant loss of public trust in nuclear energy.
- Lesson: Ambiguous or incorrect data presentation can lead even trained professionals to make disastrous decisions.

The Three Mile Island (TMI) nuclear incident, which occurred on March 28, 1979, at the Three Mile Island Nuclear Generating Station near Harrisburg, Pennsylvania, was the most serious accident in U.S. commercial nuclear power history. Though it did not result in any direct fatalities or major offsite radiation exposure, the incident led to a partial meltdown of the reactor core in Unit 2 and had profound consequences for public perception of nuclear safety, regulatory oversight, and control system design.

The failure point was rooted in a misleading control panel indicator that provided incorrect status information about a crucial safety component, a pilot-operated relief valve (PORV). During a routine maintenance event, a relatively minor malfunction in the feedwater system caused the reactor to automatically shut down. This was followed by a rise in pressure within the reactor, triggering the PORV to open as it was designed to relieve pressure. However, after the pressure normalized, the valve failed to close, allowing coolant to continuously escape from the primary system.

Here's where the critical human-machine interface breakdown occurred: the control room indicator showed the "valve closed" light, but this signal only reflected the command to close, not the actual physical status of the valve. The operators, assuming the valve had successfully closed, misinterpreted rising temperature and pressure changes as signs that the system was overfilled, not losing coolant. Acting on that incorrect assumption, they manually shut off emergency coolant systems, worsening the loss of cooling and contributing to a partial core meltdown.

Despite the presence of trained operators, alarms, and multiple system readouts, the design flaw in the data display, coupled with information overload and poor alarm prioritization, prevented the correct diagnosis of the situation. Operators were overwhelmed by hundreds of alarms, many of which were cryptic, redundant, or irrelevant to the real crisis. In this fog of information, they missed the crucial fact that the reactor was suffering from a loss-of-coolant accident (LOCA), the very scenario the safety systems were meant to prevent.

The consequence of this cascade was a partial meltdown of the reactor core, resulting in the release of small amounts of radioactive gases into the environment and a temporary evacuation advisory for the surrounding community. While no immediate injuries or deaths occurred, the psychological and political fallout was immense. The incident led to a massive erosion of public confidence in nuclear energy, sparked widespread media coverage, and triggered a nationwide halt in nuclear plant construction in the U.S. for decades.

The lesson of Three Mile Island is stark and enduring: even highly trained professionals can make disastrous decisions when faced with ambiguous, misleading, or incomplete data. The incident underscored the importance of human factors engineering in control room design, where the clarity, accuracy, and prioritization of information are just as critical as the mechanical components they monitor. It also highlighted the need for systems that report true equipment status, not just control intentions, and for alarm systems that filter and contextualize data, allowing operators to focus on what truly matters during high-stress situations.

In the aftermath, the Kemeny Commission and NRC (Nuclear Regulatory Commission) issued sweeping recommendations to improve operator training, control system design, emergency protocols, and regulatory oversight.

Many of the human-machine interface standards now used in process control and safety-critical industries can trace their roots to the insights gained from this incident.

Ultimately, Three Mile Island demonstrated that technology and training are not enough without thoughtful system design and a culture that prioritizes clear, accurate, and actionable information. In high-risk environments, how data is presented can be just as important as what the data says. When lives, infrastructure, and public trust are on the line, miscommunication even at the machine level can prove just as fatal as mechanical failure.

COVID-19 Initial Response (2020)

- Failure Point: Inaccurate early modeling and data underreporting.
- Details: Many nations relied on incomplete or suppressed infection data, and early models underestimated asymptomatic spread.
- Consequence: Delayed lockdowns, healthcare system overload, and millions of preventable deaths.
- Lesson: Data quality, transparency, and early action matter; bad data leads to delayed or wrong responses in a crisis.

The initial global response to the COVID-19 pandemic in early 2020 stands as a stark and sobering reminder of the critical importance of data quality, transparency, and timely interpretation in crisis management. Although the novel coronavirus (SARS-CoV-2) began circulating in late 2019, the world's ability to respond effectively was hampered not by a lack of tools or scientific talent but by flawed data, inaccurate modeling, and delayed recognition of the virus's true nature.

The failure point lies in several overlapping dimensions of data mismanagement and misinterpretation. First, early infection data, particularly from the outbreak epicenter in Wuhan, China, was incomplete, delayed, or actively suppressed. Initial reports underestimated both the number of infections and the virus's capability to spread before symptoms appeared. Public health officials around the world built models based on this inadequate data, assuming transmission was primarily symptomatic and that containment strategies like temperature screening and travel restrictions would be sufficient.

Second, the epidemiological models used by governments, universities, and international organizations failed to accurately reflect the role of asymptomatic and pre-symptomatic transmission, a defining feature of SARS-CoV-2 that made it far more contagious than initially believed. In many cases, national and regional models were not revised quickly enough as new evidence emerged. This led to a widespread underestimation of infection rates, particularly in the early weeks when testing was limited or unavailable.

As a result, many nations delayed lockdowns, social distancing policies, and travel restrictions, even as the virus silently spread within their borders. By the time clearer data emerged, often from overwhelmed hospitals or mass fatality events, healthcare systems were already collapsing under the strain. Ventilator shortages, ICU capacity overruns, and exhausted medical personnel became common themes across countries like Italy, Spain, the UK, Brazil, India, and the United States.

The consequences were devastating: the pandemic led to the deaths of over 7 million people globally by 2024 (according to WHO estimates), with studies suggesting that millions of those deaths could have been prevented through earlier intervention. The pandemic also exposed systemic weaknesses in global health infrastructure, including poor coordination between international health bodies, inconsistent national reporting standards, and a lack of investment in surveillance and pandemic preparedness.

The lesson of the COVID-19 initial response is deeply instructive: bad data leads to bad decisions, especially in fast-moving crises. It is not enough to model the data that is available; policymakers must also consider what data is missing, what may be misrepresented, and how evolving conditions could shift key assumptions. The early models treated COVID-19 as if it behaved like SARS (2003) or influenza, failing to account for its unique transmission dynamics. This misalignment between model and reality delayed critical public health interventions.

Equally important is the role of data transparency and trust. Suppression or distortion of early outbreak data, whether for political, economic, or image-management reasons, costs the global community precious time. It impeded the development of testing capacity, distorted case fatality rates, and prevented countries from seeing the full scope of the threat. Moreover, public trust in institutions suffered due to inconsistent messaging, shifting recommendations, and politicized data interpretation.

Finally, the pandemic underscored the importance of preemptive action based on incomplete but credible signals. Waiting for perfect data in an exponentially worsening crisis means always acting too late. Countries like South Korea, Taiwan, and New Zealand, which implemented aggressive testing, contact tracing, and border controls based on early warning signs, saw significantly better outcomes.

The early phase of the COVID-19 pandemic revealed the perils of incomplete data, rigid models, slow reaction times, and a failure to adapt as new evidence emerged. It reminded the world that in times of crisis, data must be fast, transparent, contextualized, and actionable and that leadership must be willing to act boldly before certainty is available. The next global crisis will demand not just better science but a better system for translating emerging data into decisive, lifesaving action.

Mars Climate Orbiter Loss (1999)

- Failure Point: Unit conversion error (imperial vs. metric).
- Details: One team used pound-seconds, and another used newton-seconds. The navigation software interpreted data incorrectly, causing the probe to enter Mars' atmosphere at the wrong altitude.
- Consequence: The $327 million probe was destroyed.
- Lesson: Even small mismatches in data assumptions can destroy complex missions.

The loss of NASA's Mars Climate Orbiter in 1999 remains one of the most infamous examples of how a seemingly minor data mismatch—in this case, a unit conversion error—can lead to the total failure of a highly complex scientific mission. Despite its modest footprint in public memory compared to disasters like Challenger or Chernobyl, the Orbiter's destruction offers a powerful cautionary tale in an engineering discipline, data standardization, and systems integration.

The failure point was deceptively simple: a unit conversion error between two collaborating engineering teams. One team, based at Lockheed Martin, delivered thruster performance data using imperial units, specifically pound-seconds of force (lbf·s). However, the NASA Jet Propulsion Laboratory (JPL) navigation team, which operated under the metric system as required by NASA standards, interpreted the data as newton-seconds (N·s). This mismatch introduced a systematic error in calculating trajectory corrections, causing the spacecraft to gradually drift off course.

Because the unit discrepancy was not caught during integration or software validation, the navigation software incorrectly calculated the spacecraft's trajectory through space. As a result, when the Orbiter arrived at

Mars, it entered the atmosphere at an altitude of about 57 kilometers instead of the planned 140 kilometers. This caused it to encounter atmospheric forces far greater than it was designed to withstand, and the spacecraft either burned up or skipped off into space, lost forever. The entire mission intended to study Martian climate and weather systems was destroyed, along with its $327 million price tag and countless hours of scientific labor.

The lesson of the Mars Climate Orbiter is as frustrating as it is fundamental: even the smallest data assumption, if incorrect, can lead to mission-ending failure. In this case, both engineering teams were technically correct in their calculations within their respective unit systems, but there was a fatal breakdown in cross-team communication, interface control, and systems engineering discipline. There were no automated checks to confirm units, no shared interface control document (ICD) that enforced unit consistency, and no robust cross-validation of the data's interpretation prior to critical flight operations.

NASA's investigation concluded that the failure stemmed from a lack of rigorous systems integration oversight and poor quality control in the mission's software pipeline. The assumption that both teams were working in the same unit system, an implicit expectation rather than a confirmed fact, was never verified. This highlights a critical engineering truth: never assume interoperability; verify it.

The consequences of this oversight were far-reaching. Not only did NASA lose a spacecraft and a high-value scientific mission but also the agency also faced intense scrutiny over its internal coordination, contractor oversight, and systems engineering protocols. In response, NASA revised its project management guidelines to enforce stricter interface documentation, cross-verification of assumptions, and unit consistency protocols for all future missions. This incident is now a standard case study in both aerospace and systems engineering courses as a quintessential example of "integration failure" rooted in human and organizational error, not hardware or software malfunction.

More broadly, the Mars Climate Orbiter incident underscores a universal principle: in complex, multi-team engineering projects, the smallest oversight at the data level can have cascading consequences. Whether in aerospace, medicine, nuclear energy, or manufacturing, unit mismatches, rounding errors, or inconsistent assumptions can trigger failures that far outstrip the original problem in scale and cost.

In the end, the Mars Climate Orbiter didn't fail because the science was wrong or the technology was insufficient; it failed because two smart teams never confirmed they were speaking the same numerical language. This lesson that data without context, coordination, and validation is a liability is just as urgent today as it was in 1999.

2008 Financial Crisis

- Failure Point: Overreliance on flawed risk models and incomplete credit data.
- Details: Mortgage-backed securities were rated based on models that underestimated default risk. Data about borrower income, risk factors, and debt levels was either missing or ignored.
- Consequence: Global economic collapse, millions lost homes and jobs.
- Lesson: If the model is wrong, all data that feeds it becomes misleading. Transparency and stress testing are critical.

The 2008 financial crisis was not merely an economic event; it was a catastrophic failure of data integrity, model design, and risk communication on a global scale. At the heart of the collapse were mortgage-backed securities (MBSs) and collateralized debt obligations (CDOs), complex financial instruments that promised to distribute risk across the housing market. But these instruments were built on fundamentally flawed assumptions and fed by incomplete or inaccurate data, making them ticking time bombs inside the global financial system.

The failure point centered on the overreliance on flawed risk models and the systematic omission or suppression of key credit data. Financial institutions, credit rating agencies, and investors depended heavily on quantitative models like Value at Risk (VaR) and Gaussian copula functions to assess the likelihood of default across large pools of mortgages. These models assumed that housing prices would continue to rise, that defaults would be uncorrelated, and that past performance data could predict future outcomes.

However, these models grossly underestimated the true default risk, especially for subprime mortgage loans issued to borrowers with poor credit histories, limited income verification, or high existing debt. Worse, in the push to package and sell MBSs, data about borrower income, debt-to-income ratios, and credit scores was often incomplete, outdated, or ignored entirely. Lenders, incentivized to approve as many loans as possible, relaxed underwriting standards, and originators passed risky loans off to secondary markets where true creditworthiness was obscured by aggregation.

The consequences were devastating. As housing prices began to decline and defaults surged, the risk models failed completely, revealing that these supposedly diversified investment products were far more fragile and interconnected than anyone realized. Major financial institutions collapsed, including Lehman Brothers, while others like AIG, Bear Stearns, and countless commercial banks required massive government bailouts to avoid total collapse. Credit markets froze, global trade slowed, and millions of people lost their homes, their jobs, and their savings in what became the worst financial crisis since the Great Depression.

The lesson of the 2008 financial crisis is a sobering one: when the model is wrong, even accurate data becomes misleading, and when the data itself is flawed, the illusion of precision becomes deadly. Risk models are only as good as the assumptions they embed and the completeness and accuracy of the data they consume. In this case, overconfidence in quantitative analytics blinded institutions to systemic vulnerabilities, and a widespread lack of transparency ensured that few could challenge the consensus narrative until it was too late.

Furthermore, the crisis underscored the necessity of stress testing and scenario planning. Regulators and banks failed to simulate worst-case events or to explore tail-risk scenarios, in part because the models were not designed to handle them and in part because short-term profits and market optimism overshadowed long-term prudence. When data is filtered through models that are untested, opaque, or misaligned with reality, it becomes a tool not of insight, but of delusion.

In the aftermath, financial regulations such as the Dodd-Frank Act, the Volcker Rule, and the establishment of systemic risk councils were put in place to improve transparency, enforce stricter data disclosure requirements, and require regular stress testing of financial institutions. Credit rating agencies, which had previously assigned AAA ratings to high-risk securities, were also subjected to increased scrutiny and accountability.

Ultimately, the 2008 financial crisis serves as a global case study of the perils of data complacency, overmodeling, and ignoring uncertainty. It is a powerful reminder that models should inform human judgment, not replace it, and that data without scrutiny, transparency, and systemic oversight can become a weapon of mass miscalculation.

Japanese Fukushima Daiichi Nuclear Disaster (2011)

- Failure Point: Seismic and tsunami risk models underestimated real-world conditions.
- Details: Historical data on tsunamis was not fully incorporated into reactor placement and seawall design.
- Consequence: Earthquake + tsunami disabled cooling systems, leading to core meltdowns.
- Lesson: Selective data or ignoring worst-case scenarios is a silent setup for catastrophe.

The Fukushima Daiichi nuclear disaster, triggered by the Great East Japan Earthquake and subsequent tsunami on March 11, 2011, represents one of the most significant nuclear accidents since Chernobyl and one of the clearest examples of how underestimating environmental risks and selectively using data can lead to devastating consequences. While the reactor technology functioned as designed under normal conditions, it was not prepared for the true scale of natural forces it ultimately faced, leading to a multi-reactor meltdown and a global crisis of nuclear confidence.

The failure point was rooted in flawed seismic and tsunami risk models, which underpredicted the potential magnitude of natural disasters in the region. The Fukushima Daiichi plant, operated by Tokyo Electric Power Company (TEPCO), was built based on assessments that assumed a maximum credible tsunami of approximately 5.7 meters. However, the actual tsunami that struck following the 9.0 magnitude earthquake reached heights exceeding 13 meters in some areas, over twice the expected worst case, completely overwhelming the plant's protective seawall and inundating critical infrastructure.

Crucially, historical records and geological evidence had indicated the possibility of much larger tsunamis in the region, including ancient "Jōgan tsunami" events over 1,000 years ago. But this data was not fully incorporated into design or mitigation strategies, possibly due to a combination of cost concerns, regulatory gaps, and institutional resistance to acknowledging low-probability, high-impact scenarios. TEPCO's risk models, while technically sound within their chosen parameters, ignored broader historical data that could have reshaped the assumptions underlying plant siting and safety systems.

When the tsunami hit, it disabled the backup diesel generators and electrical systems required to power the reactors' cooling systems. With no way to remove residual heat, three of the plant's six reactors experienced core meltdowns, leading to hydrogen explosions, radioactive releases, and large-scale evacuations. The contamination spread through air, soil, and water, displacing over 150,000 residents and rendering large areas uninhabitable for years. The disaster caused billions of dollars in economic losses, eroded public trust in nuclear energy, and forced Japan to shut down or reconsider most of its nuclear fleet.

The lesson of Fukushima is clear and deeply humbling: ignoring historical data, underestimating natural variability, or discounting worst-case scenarios can silently set the stage for catastrophe. The engineers and regulators involved did not fail in their technical execution; they failed in their risk framing and data inclusion. By designing for what was "likely" rather than what was plausibly possible, the system was rendered vulnerable to exactly the kind of extreme event that history had already foreshadowed.

This disaster exposed a broader issue in high-reliability fields: the tendency to rationalize away rare risks under the guise of practicality or statistical conservatism. However, in systems where the cost of failure is existential, such as nuclear power, worst-case scenario modeling is not optional; it is essential. Following the disaster, Japan's nuclear regulatory framework underwent sweeping reforms, including the establishment of the Nuclear Regulation Authority (NRA) and new guidelines requiring comprehensive seismic and tsunami risk assessments, elevated backup systems, and hardened infrastructure.

More globally, Fukushima reignited debates around risk-informed design, resilience engineering, and black swan preparedness. It underscored that true safety requires not just robust systems but robust imagination—the willingness to model and plan for uncomfortable possibilities, not just convenient probabilities.

In the end, the Fukushima disaster wasn't just about a tsunami or an earthquake; it was about data that was known but not heeded, models that were bounded by false certainty, and systems that lacked the flexibility to respond when reality defied expectations.

Y2K Readiness Panic (1999–2000)

- Failure Point: Misjudged data storage risks.
- Details: Many systems stored dates using two digits, and it was feared that clocks rolling over to "00" would cause critical systems to fail.
- Consequence: Massive, expensive global mitigation effort; very few actual problems occurred.

- Lesson: Even perceived bad data can lead to massive economic decisions. Preparedness matters, but so does accurate risk assessment.

The Y2K phenomenon, often dubbed the "Millennium Bug," emerged as a unique moment in technological and organizational history, one in which a perceived data risk prompted a massive, global response, despite relatively few actual failures. Although it did not end in disaster, the Y2K readiness panic is a critical case study in data design flaws, risk communication, economic decision-making, and the fine balance between caution and overreaction.

The failure point lay in a legacy data storage convention: for decades, computer systems stored calendar years using only two digits, "79" for 1979 and "98" for 1998, primarily to save memory in early computing environments. As a result, it was feared that when the year rolled over from "99" to "00," systems might misinterpret the date as 1900 instead of 2000. This raised alarms across every industry that relied on date-based functions, from banking, utilities, and air traffic control to military, healthcare, and infrastructure systems. The potential for cascading errors such as failed billing cycles, flawed safety routines, or total system crashes was widely feared.

What made the Y2K crisis so unique was that it was not triggered by a real-time event; it was driven by a global realization of a latent design flaw that could impact critical systems. In response, both public and private sectors initiated one of the largest preemptive IT mobilizations in history, with governments, corporations, and international agencies spending an estimated $300 to $600 billion on system audits, code rewrites, hardware replacements, and testing protocols.

By the time January 1, 2000, arrived, most critical systems had been patched, tested, or replaced, and very few actual problems occurred. A handful of minor glitches were reported (e.g., incorrect timestamps and misdated utility bills), but none caused significant service disruptions or safety incidents. To some, this confirmed the success of the mitigation effort; to others, it fueled skepticism that the threat was ever real to begin with.

The lesson of Y2K is multi-dimensional. On one hand, it demonstrated the importance of proactive risk assessment and global coordination. Arguably, the crisis was avoided because the world took the threat seriously. On the other hand, it revealed how perceived data risks, even if largely theoretical, can drive enormous economic and strategic decisions. The line between "prevented disaster" and "overhyped problem" remains blurry and debated, but the legacy is clear: preparedness matters, but so does calibrated risk assessment.

Y2K also served as a wake-up call about the unintended consequences of short-term data decisions. What began as a memory-saving practice decades earlier had grown into a potential global vulnerability, showing how technical debt can scale in silent, systemic ways. In the process, the incident brought data architecture, legacy system maintenance, and software lifecycle management into the spotlight, prompting many organizations to modernize their IT infrastructures.

Perhaps most importantly, Y2K highlighted the power of anticipation overreaction. Unlike disasters such as Fukushima or the 2008 financial crisis, where warning signs were ignored or misunderstood, the Y2K threat, though ambiguous, was widely acknowledged and addressed ahead of time. The result? A rare instance where global technological foresight may have successfully defused a looming systems failure, at the cost of considerable effort and expense.

In retrospect, the Y2K panic stands as a paradox: a non-event that reshaped global IT policy and preparedness culture, not because disaster happened, but because people feared that it might and acted in time to stop it.

7
Failure Mode and Corrective Action

Abstract
Failure analysis using the 7 M's framework (Man, Method, Machine, Material, Measurement, Mother Nature, Money) identifies root causes in diverse systems. This methodology aids in uncovering human error, process inconsistencies, equipment faults, material defects, measurement inaccuracies, environmental impacts, and financial constraints. Examples include thermocouple drift, pressure transducer errors, and more. Corrective actions are tailored for each M, ensuring comprehensive problem-solving. Implementing standardized protocols, regular calibration, training, and environmental controls enhances reliability and prevents recurrence. The 7 M's facilitate a systematic approach to identifying and mitigating failure points across industries.

Keywords: Measurement science, standards, calibration, NIST, ISO, traceability, uncertainty

Failure is a blessing. It really is! It's a key to the door to success if you are brave enough to turn it. Considering the topics of the past chapters, a plan to make sure these things do not happen again requires examination and analysis. Categorizing the cause of failure would aid in our attempt to mitigate the problems.

The 7 M's—man, method, machine, material, measurement, mother nature, and money—are a foundational framework in root cause analysis, quality management, and problem-solving methodologies such as Fishbone (Ishikawa) Diagrams, Six Sigma, and Lean Manufacturing. They provide a structured way to analyze causes of variation, failure, or inefficiency in any system. Here's the rationale behind using them:

Michael D. Holloway. Accuracy and Calibration Failures in Science and Engineering, (121–130)

Rationale for the 7 M's Framework

1. Man (People)
    - Focuses on human error, lack of training, miscommunication, or fatigue.
    - Humans operate, inspect, calibrate, and make decisions in systems.
    - Example: An operator misreads calibration instructions.
2. Method (Process)
    - Involves the standard operating procedures (SOPs), workflows, or lack thereof.
    - Process inconsistency is a major source of variation.
    - Example: No standardized procedure for recalibration frequency.
3. Machine (Equipment)
    - Refers to the performance, maintenance, and design of machinery or instruments.
    - Aging equipment, design flaws, or inadequate diagnostics can all lead to problems.
    - Example: A spectrophotometer with aging optics causing drift.
4. Material
    - Considers the inputs into the process, such as raw materials, reagents, or components.
    - Poor-quality inputs can compromise system performance or skew measurements.
    - Example: Expired buffer solution used in pH calibration.
5. Measurement
    - Includes how performance, quality, or values are evaluated or verified.
    - Calibration, instrument accuracy, and traceability fall under this.
    - Example: Using an unverified flow meter for custody transfer.
6. Mother Nature (Environment)
    - Covers external factors like humidity, temperature, dust, lighting, and vibrations.
    - Often overlooked but critically important in sensitive systems.

    - Example: Infrared thermometer readings fluctuating due to ambient air currents.
7. Money
    - Budget limitations
    - Financial constraints
    - Revaluation due to leadership refocus

Why Use the 7 M's?

- Comprehensive: Ensures no major category of influence is ignored.
- Visual: Easily maps cause in tools like Fishbone Diagrams.
- Universal: Applicable across industries from manufacturing to healthcare and metrology.
- Collaborative: Teams can tackle each "M" to gather insights from different disciplines.

In more specialized or modern contexts, an additional "M" is sometimes added to tailor the analysis:

Additional M	**Focus**
Management	Leadership, policy decisions, culture, priorities
Maintenance	Preventive or corrective maintenance of equipment
Market	External forces like demand, customer expectations
Media	Communication channels, information flow
Milieu	Workplace culture, morale, psychological safety

These additions are particularly useful in complex systems (e.g., large hospitals, regulatory labs, or aerospace manufacturing) where organizational and systemic influences are strong.

The M's provide a logic-driven structure to ensure root cause analysis isn't biased or superficial. They encourage a holistic view of a system's potential failure points, allowing for better-designed corrective actions, sustainable quality improvement, and fewer surprises during audits or crises.

7 M's Put to Use

Let's consider our failures mentioned earlier:

Thermocouple Drift in Aerospace and Power Industries

- Man: Inadequate training on thermocouple handling.
- Method: Lack of standardized calibration procedures.
- Machine: Use of outdated or worn thermocouple equipment.
- Material: Substandard thermocouple materials leading to oxidation.
- Measurement: Infrequent calibration intervals.
- Mother Nature: Exposure to extreme temperatures causing sensor degradation.

Corrective Actions:

- Implement regular calibration schedules.
- Train personnel on proper thermocouple usage. Ease. io+9Tableau+9Flagship Facility Services+9.
- Upgrade to high-quality thermocouple materials resistant to oxidation.
- Establish standardized calibration procedures.

Pressure Transducer Zero Drift (BP Oil Spill Related)

- Man: Insufficient monitoring of sensor performance.
- Method: Absence of real-time calibration checks.
- Machine: Deployment of sensors beyond their operational lifespan.
- Material: Use of materials susceptible to corrosion in marine environments.
- Measurement: Delayed detection of zero drift anomalies.
- Mother Nature: Harsh subsea conditions affecting sensor integrity.

Corrective Actions:

- Introduce continuous monitoring systems for pressure transducers.

- Schedule timely replacements of aging sensors. Output Industries | Busroot.
- Select corrosion-resistant materials for sensor construction.
- Implement real-time calibration verification protocols.

pH Electrode Calibration (Environmental Monitoring)

- Man: Improper handling leading to electrode contamination.
- Method: Inconsistent calibration techniques across operators.
- Machine: Use of electrodes beyond their recommended service life.
- Material: Degradation of electrode membranes over time.
- Measurement: Neglecting to perform routine calibration checks.
- Mother Nature: Temperature fluctuations affecting electrode response.

Corrective Actions:

- Standardize calibration procedures and provide comprehensive training.
- Replace electrodes as per manufacturer guidelines.
- Store electrodes under recommended conditions to prolong lifespan.
- Incorporate temperature compensation during measurements.

CT Scanner Dosimetry Drift (Medical Imaging)

- Man: Lack of awareness about software updates affecting dosimetry.
- Method: Inadequate protocols for post-update calibration verification.
- Machine: Aging sensors leading to inaccurate dose measurements.
- Material: Sensor materials degrading over time, affecting readings.
- Measurement: Failure to detect drift due to infrequent calibration.
- Mother Nature: Environmental factors like room temperature impacting sensor performance.

Corrective Actions:

- Implement mandatory calibration checks after software updates.
- Schedule regular maintenance and replacement of aging sensors.
- Enhance training for technicians on recognizing and addressing dosimetry drift.
- Monitor environmental conditions to ensure they remain within optimal ranges for equipment operation.

Mass Standards Drift (Pre-Kilogram Redefinition)

- Man: Handling errors leading to contamination of mass standards.
- Method: Inconsistent cleaning procedures affecting mass integrity.
- Machine: Storage equipment not maintaining required conditions.
- Material: Surface contamination altering the mass of the prototype kilogram.
- Measurement: Reliance on a single physical artifact for mass standardization.
- Mother Nature: Environmental factors causing material degradation.

Corrective Actions:

- Redefine the kilogram based on fundamental constants (e.g., Planck constant).
- Implement stringent handling and cleaning protocols.
- Upgrade storage facilities to maintain optimal environmental conditions.
- Develop and adopt alternative mass standards less susceptible to drift.

Gas Chromatograph (GC) Flame Ionization Detector (FID) Calibration Drift

- Man: Operator errors in setting detector parameters.

- Method: Irregular calibration schedules.
- Machine: Aging detectors with declining performance.
- Material: Contaminants affecting detector response.
- Measurement: Neglecting baseline drift corrections.
- Mother Nature: Ambient temperature variations impacting detector stability.

Corrective Actions:

- Establish routine calibration using certified standards.
- Train personnel on proper detector maintenance.
- Implement automated baseline correction algorithms.
- Control laboratory environmental.

Optical Emission Spectrometer (OES) Metal Analysis Drift

- Man: Inadequate cleaning and handling of electrodes and sample stands.
- Method: Lack of routine calibration or standardization using reference blocks.
- Machine: Aging arc/spark systems with inconsistent excitation energy.
- Material: Accumulation of residue from previous samples contaminating new tests.
- Measurement: Drift in baseline emission response over time.
- Mother Nature: Temperature and humidity fluctuations impacting detector performance.

Corrective Actions:

- Implement daily or shift-start calibrations using certified reference materials.
- Train operators in consistent sample prep and cleaning protocols.
- Replace aging excitation and optical components proactively.
- Use environmental control in analytical labs to stabilize readings.

Industrial Scales – Load Cell Drift

- Man: Improper loading techniques and lack of user training.

- Method: Missed recalibration intervals or non-standard calibration routines.
- Machine: Mechanical fatigue or creep in the load cell components.
- Material: Thermal expansion or contraction of materials in load cell housing.
- Measurement: Absence of zero drift detection routines or alerts.
- Mother Nature: Temperature and humidity affecting electrical resistance.

Corrective Actions:

- Adhere to ISO/IEC 17025 calibration intervals using traceable standards.
- Use automated drift compensation and self-check features.
- Implement training on proper use and routine tare/zero checks.
- Apply thermal shielding or climate control for weighing environments.

Spectrophotometer Baseline Shift in Color Measurement

- Man: Inconsistent sample presentation or mishandling of calibration tiles.
- Method: Infrequent validation or improper storage of calibration standards.
- Machine: Lamp aging, filter degradation, and detector instability.
- Material: Contamination on optical paths or calibration surfaces.
- Measurement: Drift in reflectance/transmittance baseline values.
- Mother Nature: Light and temperature variation in uncontrolled environments.

Corrective Actions:

- Perform daily verifications using certified color standards (tiles).

- Replace lamps and detectors as per preventive maintenance schedules.
- Employ dual-beam spectrophotometers for drift cancellation.
- Control ambient lighting and temperature during measurements.

Flow Meter Calibration Drift in Custody Transfer

- Man: Lack of operator understanding of flow meter performance thresholds.
- Method: Irregular validation against provers or master meters.
- Machine: Wear of turbine blades or orifice plates; fouling in sensors.
- Material: Deposits or corrosion within pipework changing flow dynamics.
- Measurement: No redundancy or comparison to secondary meter data.
- Mother Nature: Fluid temperature and pressure variations affecting readings.

Corrective Actions:

- Replace aging mechanical meters with ultrasonic or Coriolis flow meters.
- Calibrate flow meters routinely against certified flow standards.
- Use dual meters or parallel metering systems for real-time validation.
- Integrate self-diagnostic capabilities into custody transfer metering stations.

Ion Selective Electrode (ISE) Calibration for Sodium and Chloride in Water

- Man: Poor handling of sensors or inconsistent conditioning practices.
- Method: Single-point calibration or reuse of outdated standards.
- Machine: Aging electrodes and deteriorating membrane performance.

- Material: Interference from other ions (e.g., potassium and bicarbonate).
- Measurement: Lack of compensation for non-linearity at concentration extremes.
- Mother Nature: Sample temperature and ionic strength fluctuations.

Corrective Actions:

- Conduct minimum two-point calibrations using fresh certified standards.
- Establish electrode conditioning and replacement schedules.
- Use ion interference correction or shielding agents when applicable.
- Train staff on error sources and documentation of calibration history.

Infrared Thermometer Drift (Non-Contact Thermometers)

- Man: Incorrect usage (e.g., wrong distance and surface angle).
- Method: No calibration checks post-manufacturing or during field use.
- Machine: Sensor degradation, lens contamination, and faulty firmware.
- Material: Low-grade optics in mass-produced units during COVID-19.
- Measurement: Drift in temperature readings due to lack of blackbody reference.
- Mother Nature: Ambient temperature and air currents affecting readings.

Corrective Actions:

- Require routine blackbody calibration for deployed units.
- Use dual-reference or self-calibrating IR devices in critical settings.
- Provide training for correct usage and interpretation of readings.
- Implement traceable quality assurance for all deployed thermometers.

8
Universal Reasons for Failure

Abstract
Failures often result from systemic breakdowns involving flawed data, complacency, siloed decision-making, poor communication, and ignored worst-case scenarios. Flawed data originates from errors, biases, or incomplete information leading to catastrophic decisions. Complacency in models and systems fosters overconfidence, masking vulnerabilities. Siloed operations prevent cross-disciplinary validation essential for comprehensive risk assessment. Hierarchical suppression buries critical warnings from frontline experts. Ignoring worst-case scenarios leaves systems unprepared for rare but impactful events. Design flaws in interfaces mislead users, undermining decision-making. Addressing these issues requires rigorous data validation, interdisciplinary collaboration, transparency in communication, and robust risk planning.

Keywords: Quality assurance, measurement accuracy, compliance, standards, traceability, ISO 9001, verification

Across all the examples, from spaceflight to pandemics, failures stemmed not from a single point but from a systemic breakdown. The following are generalized reasons for these failures.

Flawed or Incomplete Data

Assumptions made on poor-quality, misinterpreted, or missing information.

At the root of many major technical, operational, and strategic failures lies a deceptively simple yet dangerous factor: flawed, incomplete, or misinterpreted data. When decisions, whether made by engineers, executives, algorithms, or policymakers, are built on assumptions drawn from poor-quality, missing, or context-deficient data, the resulting conclusions can be not only wrong but catastrophically so.

Michael D. Holloway. Accuracy and Calibration Failures in Science and Engineering, (131–142)

Flawed data may arise from faulty sensors, data entry errors, biased collection methods, or corrupted files. Incomplete data can result from gaps in measurement, unrecorded variables, lack of historical context, or missing edge-case scenarios. Even when data is technically accurate, it can be misinterpreted due to poor communication, oversimplified models, or cognitive bias.

The effects of these issues are evident across industries and crises:

- In the Challenger disaster, temperature-related O-ring data existed but was not clearly communicated or statistically supported, leading to a fatal decision under pressure.
- During the COVID-19 pandemic, early models underestimated asymptomatic spread due to lack of data, resulting in delayed containment measures and overwhelmed healthcare systems.
- The 2008 financial crisis unfolded as risk models, fed by incomplete borrower data, grossly misjudged default risk, turning structured securities into instruments of global collapse.
- Even the Mars Climate Orbiter was lost due to a unit mismatch, not because the data was wrong, but because the assumption behind its interpretation was.

These examples reinforce a critical lesson: Assumptions fill the voids that flawed or incomplete data creates. When left unchecked, these assumptions become the basis for automated decisions, operational protocols, and crisis responses, amplifying their potential to cause harm.

To mitigate this risk, organizations must:

- Validate data quality and completeness at every stage of the decision-making chain.
- Treat missing data as a risk factor, not an invitation to assume.
- Encourage cross-functional communication to clarify assumptions and share contextual knowledge.
- Use sensitivity analyses and scenario planning to explore what happens when data assumptions are wrong.
- Establish a culture of metrological and analytical humility, recognizing that certainty in flawed data can be far more dangerous than uncertainty acknowledged and addressed.

Bad data quietly undermines good intentions. Preventing that means building systems and cultures that do not just demand data but question it, test it, and understand its limitations before trusting it to shape critical decisions.

Complacency or Overconfidence in Models/Systems

Trust in established procedures or software without questioning edge cases or anomalies.

In high-reliability environments, where precision and consistency are essential, there's a subtle but powerful risk that creeps in over time: complacency. When operators, engineers, or decision-makers place too much trust in models, procedures, or software systems without questioning edge cases, anomalies, or underlying assumptions, they create a false sense of security that can mask growing vulnerabilities. Overconfidence in established systems does not prevent failure; it makes it more likely to be missed until it is too late.

This mindset often develops when systems perform reliably for extended periods. Familiarity breeds comfort, and comfort breeds assumption: that procedures are foolproof, that models are universally valid, and that "nothing bad has happened so far, so nothing will." But real-world events show that unchallenged trust in automated systems or historical performance can set the stage for catastrophic surprises.

Examples of this failure mode abound:

- In the Three Mile Island nuclear incident, operators trusted the control panel indicators, not realizing they displayed valve command status, not the actual position. This assumption led to misdiagnosing a loss-of-coolant event and a partial core meltdown.
- During the 2008 financial crisis, risk models like Value at Risk (VaR) and the Gaussian copula were treated as infallible. Decision-makers failed to question how these models would behave under extreme, correlated defaults leading to the collapse of major financial institutions.
- The Fukushima disaster occurred in part because TEPCO and regulatory bodies trusted historical models that excluded evidence of rare but massive tsunamis, leading to underdesigned sea walls and cooling system vulnerabilities.

- In autonomous vehicle development, several fatal crashes have been linked to driver overreliance on driver-assist systems, where human operators failed to intervene because they had grown too confident in the technology's judgment.

At the heart of these failures is a shared pattern: systems drift from being tools to being crutches. Procedures meant to enhance human oversight gradually replace it. Models designed to assist in decision-making become unquestioned oracles, and anomalies are dismissed as noise rather than signals.

To counter this, organizations must embed deliberate skepticism into their operational culture:

- Continuously test models against edge cases, not just average performance.
- Encourage challenging assumptions, especially those built into automation or long-standing processes.
- Foster "what if" thinking in safety reviews, audits, and system validations.
- Create fail-safe mechanisms that prompt human intervention when models behave outside expected ranges.
- Train staff not just in procedures, but in understanding the intent, limitations, and failure modes of those procedures.

The paradox is this: the more reliable a system becomes, the more dangerous complacency around it becomes. The key is not to distrust systems but to respect them enough to question them, especially when they appear to be running smoothly. Because often, disaster does not start with chaos; it starts with quiet, unchallenged confidence.

Lack of Cross-Disciplinary Validation

Decisions are made in silos, without consulting alternative perspectives or methods.

In complex systems, whether technological, medical, financial, or environmental, no single discipline holds all the answers. Yet time and again, critical decisions are made within silos, with experts working in isolation from one another and without the benefit of alternative perspectives or validation methods. This lack of cross-disciplinary validation often leads to blind spots, flawed assumptions, and overlooked failure modes because the right questions were never asked by the right people at the right time.

At its core, this issue is not a failure of intelligence or effort, but of collaboration and integration. Engineers may solve for technical performance without consulting human factors experts. Data scientists may model risk without input from domain specialists who understand real-world context. Medical researchers might focus on clinical outcomes while ignoring social, behavioral, or logistical factors that affect implementation. These gaps allow systems to drift into dangerous territory, not through malice, but through myopia.

History offers striking examples of what happens when decisions lack interdisciplinary scrutiny:

- In the Challenger disaster, engineers raised concerns about O-ring performance, but the significance of their data was not effectively communicated to NASA managers, many of whom lacked the engineering background to interpret the risk. The organizational decision-making process failed to integrate technical dissent into policy.
- The Mars Climate Orbiter was lost due to a unit mismatch between two engineering teams, one using imperial units and the other metric. Had systems engineers, software developers, and navigators been engaged in deeper cross-checking and interface validation, the conversion error could have been caught before launch.
- The COVID-19 pandemic response highlighted the dangers of siloed modeling: early public health strategies focused on virology and epidemiology but often neglected logistics, behavioral psychology, or socioeconomic impact, resulting in policies that were technically sound but practically flawed or inconsistently adopted.
- In the 2008 financial crisis, quantitative risk analysts used mathematical models to assess mortgage-backed securities without integrating insights from real estate experts, behavioral economists, or those familiar with the actual conditions of the mortgage market. As a result, critical signs of instability were ignored.

The lesson is clear: disciplinary expertise is essential, but alone, it is not enough. Complex systems require a diversity of perspectives to fully understand their behavior, stress points, and potential failure modes. Without this, solutions become fragile, and early warnings are missed.

To counteract siloed thinking, organizations must:

- Integrate cross-functional teams in decision-making and system design, including not only technical experts but also operational, human factors, legal, and cultural voices.
- Encourage devil's advocacy and "red teaming" from adjacent disciplines to test plans and models.
- Require validation through independent or orthogonal methods, e.g., verifying numerical models with field tests or supplementing AI predictions with domain expert review.
- Break down organizational barriers by fostering a culture of mutual respect across disciplines, where questions are welcomed rather than dismissed.

Ultimately, cross-disciplinary validation is not a luxury; it is a safeguard against the blind spots of specialization. The most resilient systems are not built by the smartest individuals, but by diverse teams that challenge each other to see the full picture. In a world where complexity is the norm, collaboration is not just a best practice; it is a survival strategy.

Poor Communication and Hierarchy Suppression

Engineers, operators, or data scientists raised red flags that were ignored or lost in bureaucracy.

One of the most dangerous and recurring contributors to catastrophic failures across industries is not technical malfunction or flawed data; it is poor communication within hierarchical systems, especially when warnings from front-line experts are ignored, dismissed, or suppressed. When organizations prioritize authority, reputation, or deadlines over open dialogue and dissent, they create conditions where vital red flags are buried beneath bureaucracy, and the people closest to the problem are the least empowered to influence the solution.

This phenomenon, known as hierarchy suppression, occurs when individuals at lower levels—engineers, operators, technicians, or data analysts—detect anomalies or foresee risks but lack the authority, platform, or institutional support to escalate their concerns. Their voices are either filtered out by management layers, dismissed due to a "lack of big-picture understanding," or stifled by a culture where challenging senior decisions are discouraged.

There are numerous tragic examples of this failure mode:

- In the Challenger space shuttle disaster, engineers from Morton Thiokol explicitly warned that the O-rings would not perform reliably at low temperatures. But when NASA managers pushed for a green light, senior leaders at Thiokol reversed the engineers' recommendation, choosing managerial pressure over technical truth, with fatal consequences.
- At Three Mile Island, control room operators tried to interpret conflicting data amidst a confusing alarm system, but the hierarchical structure of communication delayed effective response and failed to adapt procedures to real-time insights from the front lines.
- The Fukushima disaster also included early warnings about tsunami risk backed by historical data that were voiced by researchers and junior regulatory officials but were downplayed by utility executives and regulators who prioritized cost and feasibility over low-frequency threats.
- In the 2020 COVID-19 pandemic, data analysts and epidemiologists in some countries flagged alarming patterns of spread weeks before national lockdowns were issued, but their warnings were filtered, delayed, or disregarded due to political or institutional inertia.

These failures are not just about communication; they are about organizational culture and power dynamics. In many systems, especially those that are risk-averse or status-driven, individuals are rewarded for compliance over courage. Speaking up becomes risky. Questioning leadership becomes taboo. Worst of all, decisions are made in echo chambers devoid of dissenting voices that could save lives, projects, or reputations.

To prevent this, organizations must actively dismantle hierarchy suppression and foster a culture of psychological safety and operational transparency:

- Establish formal escalation paths that allow technical staff to raise concerns directly to senior leadership, bypassing mid-level filters when necessary.

- Encourage anonymous issue reporting and post-mortem analysis where all ranks are invited to share observations and contradictions.
- Train managers to listen actively and humbly, rewarding those who surface problems early even when the message is uncomfortable or inconvenient.
- Implement "stop work authority" policies, common in high-risk industries, empowering any employee to halt operations if safety or integrity is compromised.
- Normalize dissent by celebrating cases where pushing back prevented error, rather than only rewarding on-time, on-budget compliance.

Ultimately, the cost of ignoring a well-founded warning is far greater than the cost of investigating a false alarm. A resilient system is one that listens before it breaks because, in complex operations, the people who know what's going wrong are often the ones furthest from the boardroom. The challenge for leaders is to make sure those voices are not just heard but heeded.

Failure to Anticipate or Simulate Worst-Case Scenarios

Risk assessments excluded low-probability, high-impact events.
In complex systems, the most devastating failures often arise not from ordinary malfunctions but from extraordinary events that were never seriously considered. When organizations exclude worst-case scenarios from their risk assessments, especially those deemed too unlikely, too uncomfortable, or too costly to address, they create blind spots that can ultimately prove fatal. These low-probability, high-impact events (also known as "black swans" or "tail risks") are rare by definition, but history shows they are also predictable in hindsight and devastating in outcome.

The failure to simulate extreme scenarios stems from several systemic tendencies: overreliance on historical data, optimism bias, political or economic pressure to understate risk, and the natural human discomfort with imagining catastrophic failure. But as case after case has shown, excluding the improbable is no guarantee it will not happen; it just ensures you are unprepared when it does.

Consider some of the most infamous examples:

- At Fukushima Daiichi (2011), historical records of ancient mega-tsunamis were known but omitted from design parameters. The reactor seawalls were built for a 5.7-meter wave; the actual tsunami exceeded 13 meters. The result: core meltdowns, mass evacuations, and a long-term nuclear crisis, all because worst-case data was deemed implausible.
- The Challenger disaster (1986) was preceded by multiple warnings about O-ring erosion at low temperatures. Yet mission planning did not incorporate the possibility of temperature-induced seal failure, even though it was well within the realm of physical plausibility.
- In the 2008 financial crisis, risk models failed to account for the system-wide collapse of housing markets and the possibility that mortgage defaults could be highly correlated. The assumed worst case was merely a dip, not a chain reaction that would bring down global banks.
- The Deepwater Horizon oil spill (2010) occurred after BP and its contractors failed to adequately simulate a blowout scenario at extreme depths. While a worst-case discharge volume estimate existed internally, it was not disclosed to regulators and not incorporated into emergency preparedness plans.

In each of these cases, the systems themselves were technically advanced, but the imagination behind their risk models was limited. The unthinkable was not unthinkable; it was just unbudgeted, unsimulated, or bureaucratically inconvenient.

To build resilient systems, organizations must adopt a "worst-reasonable-case" mindset, one that not only acknowledges rare events but actively plans for them:

- Stress testing and scenario analysis should go beyond average-case simulations and explore edge conditions: cascading failures, resource shortages, dual-mode failures, and long-term disruptions.
- Cross-functional scenario planning teams should include skeptics, systems thinkers, and domain specialists who can challenge base-case assumptions.

- Red-teaming and pre-mortem exercises should be routine in high-risk industries, allowing organizations to simulate failure before it happens.
- Risk assessments must account not just for likelihood, but for consequence severity, recognizing that a 1% chance of catastrophe still warrants mitigation if the stakes are high enough.

Most importantly, leadership must be willing to listen when the simulations show something they do not want to see. Often, worst-case scenarios are excluded not because they're implausible but because acknowledging them would require harder decisions, greater investment, or delays.

In a world of increasingly interconnected systems—climate, technology, infrastructure, and finance—the true danger lies not in unlikely events but in unpreparedness for them. A failure to simulate the worst is not just a failure of imagination; it's a failure of responsibility. Because when the stakes are measured in lives, livelihoods, or global stability, it is not enough to plan for what's likely; we must prepare for what's possible.

Interface and Design Flaws

Data are presented in misleading ways, or systems are unable to self-report when calibration or assumptions drift.

In the modern era of complex, data-driven operations, how information is presented is just as critical as the data itself. Even accurate readings can lead to catastrophic outcomes if they are displayed in a misleading way, misaligned with user expectations, or buried in cluttered interfaces. Just as dangerous are systems that fail to detect or communicate when their underlying assumptions or calibrations have drifted. These interface and design flaws turn tools meant to ensure safety and precision into silent contributors to failure.

The core issue is that many systems are designed with functionality in mind, not clarity. Indicators show a value but not whether that value is valid. Alerts trigger but without prioritization or context. Or worse, systems continue to report as "normal" even when internal conditions no longer support that status. These are not technical failures alone; they are failures of human-machine interaction, and they often manifest in high-stakes environments where time, comprehension, and trust are limited resources.

Several historical incidents illustrate the high cost of these flaws:

- At Three Mile Island (1979), a misleading control room indicator showed that a relief valve had been "closed," when in reality, it had only received the command to close, not confirmation that it had done so. Operators, trusting the display, misdiagnosed the problem and shut off emergency cooling, contributing to a partial core meltdown.
- The Challenger disaster also involved interface shortcomings. Engineers were looking at fragmented and inconsistently formatted temperature and erosion data, but without a clear visual correlation between O-ring performance and cold weather, the data failed to compel decision-makers to delay the launch.
- In the CT scanner dosimetry drift crisis (2008–2012), faulty calibration combined with user interfaces that did not visibly reflect software updates led to patients receiving excess radiation doses. The system reported "normal" values, but no visible cue indicated that the dose model had changed after software revisions.
- Many consumer-grade infrared thermometers during the COVID-19 pandemic continued to show normal body temperature readings, even as sensor drift, ambient interference, or calibration loss made those readings unreliable. There were no alerts to prompt recalibration, allowing flawed data to guide critical screening decisions.

These examples highlight a dangerous principle: a calm, "normal" display is not the same as a healthy system. When interfaces fail to report their own confidence, or when users cannot easily distinguish between actual values and assumed or estimated ones, trust becomes misplaced. Systems must not only show what they know but also how well they know it.

To address these risks, systems must be designed with situational awareness, cognitive ergonomics, and self-monitoring in mind:

- User interfaces should reflect system health, not just output values. If a sensor is drifting or a model is out of date, visual cues or warning states should alert users before decisions are made.

- Implement status-aware indicators, for example, "Valve commanded to close" vs. "Valve verified closed," to eliminate ambiguity.
- Integrate self-calibration or calibration tracking systems that warn users when internal consistency checks fail or time-based recalibration is overdue.
- Avoid information overload by prioritizing and filtering alarms. Group-related alerts suppress known noise signals and draw attention to outliers and emerging patterns, not just individual data points.
- Design interfaces using human-centered design principles, test them under stress, incorporate diverse user feedback, and simulate edge cases to evaluate usability under real conditions.

Ultimately, interfaces are more than dashboards; they are decision environments. If a system miscommunicates its own status or conceals critical warnings behind false confidence, it silently enables human error. And when lives, infrastructure, or trust are at stake, a clean-looking screen can be the most dangerous thing of all.

In resilient systems, design is not just about aesthetics or efficiency; it is about truthfulness, transparency, and trust. When the interface lies, even unintentionally, the data becomes dangerous and the operator becomes blind.

9

Building Technical Safeguards

Abstract
Safeguards against calibration disruptions involve multiple strategies. NIST-certified SRMs ensure traceability and consistency. ISO 11171:2016 protocols standardize calibration and update processes. Alerts and bulletins inform stakeholders of changes. Multi-modal validation cross-verifies results, reducing reliance on a single method. Calibration logs track standards and versions used. Accredited lab practices enforce quality and traceability. Certification and training updates build awareness of evolving standards. These measures collectively enhance measurement reliability, ensuring data integrity and informed decision-making in critical applications.

Keywords: Innovation, prototyping, reliability, performance, testing, validation, optimization

Since the SRM 2806 calibration shift incident, multiple safeguards have been implemented to prevent future undetected drift or disruptive changes in particle count calibration. Here's what is now in place:

NIST Traceability with Enhanced Oversight

New SRMs (2806a/b/c) are now:

- Statistically validated using multiple techniques (SEM, flow imaging, optical).
- Issued with uncertainty budgets and SI traceability.
- Batch-certified and reproducible, ensuring tighter lot-to-lot consistency.

In the aftermath of the SRM 2806 calibration shift, the metrology and reliability communities recognized the need for more robust controls to prevent similar disruptions in the future. As a result, several safeguards have been implemented to ensure greater transparency, accuracy, and traceability in particle count calibration. Foremost among these are enhanced NIST

Michael D. Holloway. Accuracy and Calibration Failures in Science and Engineering, (143–152)

oversight and traceability protocols applied to all subsequent Standard Reference Materials, including SRM 2806a, 2806b, and 2806c.

These newer SRMs are now subjected to statistical validation using a combination of advanced techniques, including scanning electron microscopy (SEM), flow imaging, and optical sizing methods. This multi-modal approach ensures that the particle size distributions are accurately characterized and cross-verified, minimizing the risk of misclassification or drift due to reliance on a single measurement technique.

Each SRM is also released with a full uncertainty budget and explicit traceability to SI units, providing end-users with a clearer understanding of measurement confidence and allowing laboratories to quantify the precision of their particle count results. In addition, these reference materials are batch-certified and designed for reproducibility, which ensures tight lot-to-lot consistency, a major improvement over the variability seen in earlier test dust formulations. These metrological upgrades lay the groundwork for higher confidence in cleanliness assessments and foster more reliable decision-making across maintenance, warranty, and quality assurance programs.

Updated ISO Standards with Change Management

ISO 11171:2016 now includes:

- Detailed protocols for size verification and optical calibration.
- Requirements for documenting calibration changes.
- Procedures for validating instrument response after calibration shifts.

A second major safeguard implemented to prevent future calibration-related disruptions is the introduction of updated ISO standards with formal change management procedures, most notably reflected in ISO 11171:2016. This revision represents a significant advancement in how optical particle counters (OPCs) are calibrated, verified, and maintained.

ISO 11171:2016 now includes detailed protocols for particle size verification and optical system calibration, providing clearer, more standardized procedures for ensuring that particle sizing is consistent and accurate across different instruments and laboratories. These protocols reduce ambiguity in edge detection and sizing interpretation, two of the primary weaknesses in earlier calibration practices.

Crucially, the updated standard also includes mandatory requirements for documenting calibration changes. This includes maintaining calibration records, noting shifts in size thresholds or traceability references, and ensuring that any updates are clearly flagged for comparison with historical data. These documentation practices are essential for long-term trend analysis and for maintaining the integrity of cleanliness monitoring programs.

Moreover, ISO 11171:2016 introduces procedures for validating instrument response after a calibration shift, ensuring that devices are not only recalibrated properly but also tested to confirm that their performance aligns with the updated standards. This validation step provides an added layer of assurance that particle count results remain accurate and interpretable even after significant metrological changes. Collectively, these changes mark a shift toward a more transparent, disciplined, and traceable calibration environment.

Industry Alerts & Historical Data Correction Guidelines

Organizations like NFPA, ASTM, STLE, and ICML now issue:

- Change alerts when standards or SRMs are updated.
- Recommendations to flag or normalize data before/after calibration shifts.
- Guidance on archiving and annotating legacy particle count data.

A third major safeguard now helping to prevent future disruptions from calibration shifts is the implementation of industry alerts and historical data correction guidelines. Leading organizations such as NFPA, ASTM, STLE, and ICML have recognized the importance of clear communication and data integrity when metrological standards evolve.

These organizations now regularly issue change alerts to notify stakeholders when standards, SRMs, or calibration procedures are updated. These alerts are disseminated across technical bulletins, certification programs, and professional networks, ensuring that changes are recognized not only by laboratory professionals but also by maintenance teams, OEMs, and reliability engineers in the field.

In addition to alerts, these groups provide practical recommendations for flagging and normalizing data collected before and after calibration changes. This ensures that trend analyses and cleanliness benchmarks remain valid and comparable, even across historical transitions in particle sizing protocols. Such normalization is crucial to avoid false interpretations of contamination trends or system degradation.

Furthermore, they offer guidance on archiving and annotating legacy particle count data, emphasizing the need to retain historical records in context. By labeling data according to the calibration standard in use at the time of measurement, organizations can preserve analytical continuity and avoid the common pitfall of comparing incompatible data sets. This structured approach to data management strengthens the long-term reliability of contamination monitoring programs and enhances accountability in asset health diagnostics.

Multi-Modal Particle Sizing Validation

Labs and OEMs are encouraged to:

- Cross-check particle counts with imaging systems (like SEM or flow-imaging cameras).
- Perform inter-lab comparisons (round-robin testing) for validation.
- Use redundant measurement techniques in critical applications (e.g., military and aerospace).

A fourth safeguard now widely adopted to improve calibration reliability is the practice of multi-modal particle sizing validation. Recognizing that no single method is immune to bias or limitations, laboratories and OEMs are now strongly encouraged to validate particle count data through multiple complementary techniques.

One key recommendation is to cross-check optical particle counter results with imaging systems, such as scanning electron microscopy (SEM) or flow-imaging cameras. These technologies offer direct visualization and precise measurement of particle dimensions and morphology, allowing users to verify the accuracy of size classifications produced by light-scattering instruments.

Additionally, inter-laboratory comparisons, commonly known as round-robin testing, are promoted to benchmark instrument performance across different facilities. By comparing results on identical samples using similar methods, labs can identify discrepancies, verify calibration accuracy, and maintain consistency within a broader industry context.

For critical applications in sectors like military, aerospace, and nuclear, it's now best practice to employ redundant measurement techniques. This may include parallel use of optical, gravimetric, and imaging-based particle analysis to ensure cross-validation and eliminate blind spots. These multi-modal approaches increase confidence in contamination control programs and provide a more resilient foundation for making high-stakes reliability and maintenance decisions.

Instrument Design with Self-Validation Features

Modern particle counters now include:

- Self-check calibration routines with internal references.
- Ability to log and time-stamp calibration versions used.
- Warnings or errors if calibration is outdated or deviates from SRM specs.

The fifth and perhaps most technologically advanced safeguard is the integration of self-validation features into modern particle counters. These innovations are designed to provide real-time assurance that instruments are operating within calibrated parameters, helping users detect and correct drift or outdated settings before data integrity is compromised.

Contemporary particle counters now include self-check calibration routines, which rely on internal reference standards to verify performance. These automated checks can be scheduled periodically or triggered manually, allowing users to confirm that the instrument's sizing and counting functions remain within specification without needing to remove the unit from service for recalibration.

Another key feature is the ability to log and time-stamp the calibration version used during each measurement cycle. This creates a clear digital audit trail, enabling users to trace particle count data back to specific calibration standards (e.g., SRM 2806a vs. 2806c), which is critical for maintaining the traceability and credibility of long-term cleanliness records.

Additionally, many instruments now issue automated warnings or error messages if calibration settings are outdated, missing, or deviate from the certified SRM specifications. These built-in safeguards act as a frontline defense, preventing miscalibrated instruments from producing unreliable data, a capability especially valuable in high-reliability environments like aviation, power generation, and advanced manufacturing.

Together, these design enhancements transform the particle counter from a passive measurement tool into an active partner in quality assurance, helping users maintain confidence in their data and supporting a more rigorous and transparent calibration ecosystem.

Regulatory and Quality System Requirements

Accredited labs under ISO/IEC 17025 or OEM quality systems are now required to:

- Document traceability to current SRMs.
- Track calibration revision histories.
- Train personnel on metrological traceability and standard updates.

A sixth and essential safeguard reinforcing calibration integrity is the implementation of regulatory and quality system requirements, particularly within laboratories and organizations operating under ISO/IEC 17025 accreditation or formal OEM quality management systems. These frameworks now mandate a higher level of rigor in the handling of calibration standards and data traceability.

Accredited labs are explicitly required to document traceability to current Standard Reference Materials (SRMs), ensuring that all particle count measurements can be linked to recognized, SI-traceable references such as SRM 2806a, 2806b, or 2806c. This requirement strengthens accountability and enables external auditors to verify the scientific basis for reported results.

In addition, labs and OEMs must now track calibration revision histories, maintaining detailed records of when calibration changes occurred, which SRMs or procedures were in use at specific times, and how these changes may impact measurement comparability. This allows for accurate benchmarking, facilitates historical data correction, and supports defensible reporting in regulated industries.

Crucially, these quality systems also require ongoing training of personnel in metrological traceability and standard updates. Staff involved in particle counting must be equipped to recognize when calibration standards change, understand the implications for data interpretation, and apply appropriate corrective actions. By embedding calibration literacy into formal quality and compliance processes, organizations create a more resilient infrastructure that minimizes the risk of undetected drift or misinformed decisions in contamination control and asset reliability.

Education and Certification Requirements

- Certification bodies (e.g., ICML, STLE, Noria, and CLS) now include:
- Curriculum on ISO 4406, ISO 11171, and SRM impacts.
- Case studies of the SRM 2806 shift and how to manage similar changes.
- Emphasis on interpreting particle count data in context.

The seventh and final safeguard ensuring resilience against future calibration disruptions is the strengthening of education and certification requirements across the lubrication, reliability, and condition monitoring professions. Leading certification bodies such as ICML (International Council for Machinery Lubrication), STLE (Society of Tribologists and Lubrication Engineers), Noria Corporation, and the Certified Lubrication Specialist (CLS) program have all expanded their curricula to address the critical role of calibration standards and traceability in data interpretation.

Modern training programs now include comprehensive instruction on ISO 4406 (the particle cleanliness coding system), ISO 11171 (the calibration standard for liquid particle counters), and the impact of SRMs such as 2806a, 2806b, and 2806c. This ensures that certified professionals are not only familiar with the standards themselves but also understand the metrological foundations behind them.

Importantly, these programs feature case studies of the SRM 2806 calibration shift, using it as a real-world example of how seemingly small changes in reference materials can lead to widespread misinterpretation if not properly managed. These case studies help professionals build critical thinking skills and apply structured responses when facing similar changes in the future.

There is also a strong emphasis on interpreting particle count data within the correct context, considering calibration versions, instrument limitations, historical baselines, and system dynamics. By embedding these principles into certification requirements, the industry is fostering a more informed workforce, one that can make sound decisions rooted in both technical knowledge and calibration awareness. This educational evolution is key to preventing the recurrence of past mistakes and advancing the overall reliability of fluid cleanliness monitoring.

Safeguards Against Future Calibration Disruptions

To maintain calibration integrity and prevent undetected drift, organizations employ a series of safeguards that each serve a distinct but interrelated purpose. NIST-certified Standard Reference Materials (SRMs) ensure traceability and consistency across measurements by anchoring results to a recognized national standard. ISO 11171:2016 protocols provide a structured framework for calibration, enabling standardized update procedures and harmonized methods across labs and facilities. Alerts and industry bulletins play a critical role in keeping stakeholders informed about changes in calibration standards, emerging risks, or revisions to best practices. Multi-modal validation: the use of more than one calibration or verification method adds a layer of redundancy, minimizing overreliance on a single approach and improving confidence in results. Instrument calibration logs track which standards and software versions were in use at the time of calibration, helping reconstruct issues if measurement discrepancies arise. Accredited laboratory practices, often aligned with ISO/IEC 17025, guarantee that calibration activities meet rigorous quality assurance benchmarks and are traceable. Finally, ongoing certification and training updates reinforce awareness among technicians and engineers, ensuring they stay aligned with evolving technical standards and industry expectations. Together, these safeguards create a robust system of checks and transparency that supports reliable, repeatable, and defensible measurement practices.

Safeguard	**Purpose**
NIST-certified SRMs with traceability	Ensures consistency and transparency
ISO 11171:2016 protocols	Standardized calibration and update processes
Alerts and industry bulletins	Keep stakeholders informed

(*Continued*)

(*Continued*)

Safeguard	Purpose
Multi-modal validation	Cross-verification prevents sole dependency on one method
Instrument calibration logs	Tracks which standard/version is in use
Accredited lab practices	Enforces quality and calibration traceability
Certification & training updates	Builds awareness of evolving standards

10

Summation and Systemic Safeguards

Abstract
Safeguards against calibration disruptions involve multiple strategies. NIST-certified SRMs ensure traceability and consistency. ISO 11171:2016 protocols standardize calibration and update processes. Alerts and bulletins inform stakeholders of changes. Multi-modal validation cross-verifies results, reducing reliance on a single method. Calibration logs track standards and versions used. Accredited lab practices enforce quality and traceability. Certification and training updates build awareness of evolving standards. These measures collectively enhance measurement reliability, ensuring data integrity and informed decision-making in critical applications.

***Keywords*:** Certification, audits, legal standards, documentation, inspection, conformity, enforcement

Fundamentally, the overall solution to safeguard against calibration-related incidents like unnoticed shifts in standards or instrument drift comes down to establishing a robust metrological infrastructure grounded in traceability, transparency, and vigilance.

Here's a breakdown of core principles and practices that form a systemic defense:

Metrological Traceability to SI Units

- Definition: Every measurement should be traceable to a recognized standard unit (SI) through an unbroken chain of comparisons, each with known uncertainties.
- Why it matters: Traceability ensures consistency across time, equipment, and location, which is critical when calibrations change.

Michael D. Holloway. Accuracy and Calibration Failures in Science and Engineering, (153–158)

- Implementation: Use calibration labs accredited to ISO/IEC 17025, referencing national/international standards (e.g., NIST and PTB).

At the foundation of any reliable measurement system is metrological traceability, the principle that every measurement result must be linked to a recognized standard unit, specifically an SI unit, through an unbroken chain of comparisons, each accompanied by a known and documented measurement uncertainty.

Why It Matters
Traceability is what allows measurements taken today to be compared confidently with those taken years ago, on different equipment, in different locations. It provides the backbone of measurement consistency, especially vital when calibration standards evolve (as seen in cases like SRM 2806a or the kilogram redefinition). Without traceability, changes in instrument readings cannot be reliably interpreted, undermining long-term trend analysis, quality control, regulatory compliance, and safety.

Implementation
To achieve traceability, organizations must work with calibration laboratories accredited to ISO/IEC 17025, which ensures that calibration methods, equipment, and reference standards are all rigorously controlled and audited. These labs reference national metrology institutes like NIST (USA), PTB (Germany), or LNE (France), which maintain primary standards tied directly to the SI system. Documentation of each link in the calibration chain, including uncertainty budgets, is essential for maintaining traceability throughout the measurement process.

This principle is not just theoretical; it underpins trust in every data point, from particle counts and flow rates to drug dosages and material certifications.

Standard Change Management Process

Problem: Changes to reference standards (e.g., SRM 2806a) often went uncommunicated to field users.

Solution: Adopt a formal Standard Change Notification (SCN) protocol:

- Alert stakeholders at all levels (OEMs, labs, technicians, auditors).

- Include technical impact statements and transitional guidelines (e.g., how to compare old and new data).
- Timestamp and version-tag calibration data for traceability.

One of the most disruptive vulnerabilities in calibration systems has historically been the lack of structured communication when reference standards change, a problem starkly illustrated by the transition from SRM 2806 to SRM 2806a, where many field users were unaware that particle sizing benchmarks had shifted. To prevent such costly oversights, organizations must implement a formal Standard Change Notification (SCN) protocol as part of their systemic metrology defense.

Problem: When changes to calibration standards or reference materials occur, whether due to recharacterization, redefinition, or improved traceability, they often fail to reach frontline users, such as lab technicians, reliability engineers, or field auditors. This results in misinterpreted data, false trends, and decision-making based on incompatible measurement baselines.
Solution: Standard Change Notification (SCN) Protocol
A formal SCN protocol ensures transparency, traceability, and continuity when metrological standards evolve. Key elements include:

- Multi-tiered stakeholder alerting: Immediately notify OEMs, calibration labs, end-users, quality managers, and auditors when a standard is updated or replaced. Communications should use redundant channels (e.g., bulletins, training platforms, and direct notifications).
- Technical impact statements: Clearly articulate what changed, why it matters, and how measurement outputs will be affected. Include guidance on comparing legacy data with new results, especially if thresholds, tolerances, or interpretations will shift.
- Version tagging and timestamping of calibration data: Every calibration record should be version-controlled, with explicit reference to the standard and method used (e.g., "Calibrated using SRM 2806a, ISO 11171:2016, traceable to NIST"). This allows users to reconstruct measurement history and maintain analytical continuity across transitions.

Adopting a standardized SCN protocol transforms change from a potential liability into a planned, documented, and interpretable evolution, ensuring that every calibration shift enhances accuracy without sacrificing trust.

Calibration Lifecycle Management

- Move beyond single-point calibration
- Use calibration curves, ongoing verification, and automated diagnostics
- Track:
- Calibration due dates
- Drift logs
- Instrument firmware/software changes that may alter performance

Traditional approaches to calibration have often relied on single-point events, a one-time adjustment, or checks performed at scheduled intervals. However, in today's precision-driven, compliance-sensitive environments, this is no longer sufficient. Organizations must shift to Calibration Lifecycle Management: a comprehensive, continuous process that accounts for performance over time, not just at a single moment.

Why It Matters
Modern instruments can drift between calibrations, experience performance changes after firmware or software updates, or react to environmental and operational stressors. A lifecycle approach ensures that calibration is treated as a living process actively monitored, documented, and evaluated across the instrument's operational lifespan.

Key Practices:

- Use calibration curves instead of relying on single-point checks. By generating a curve across multiple standard points, users can verify linearity and detect nonlinear drift before it impacts results.
- Implement ongoing verification routines using secondary standards or built-in references to confirm calibration stability between formal recalibration cycles. Many modern

instruments offer automated diagnostic checks to flag anomalies in real time.

- Maintain a centralized system to track:
  - Calibration due dates, ensuring timely recertification and reducing downtime risks.
  - Drift logs, recording deviations found during verification or recalibration. This data helps identify instruments that are degrading or operating outside of expected performance bounds.
  - Firmware/software changes, which can silently impact how sensors interpret signals or report values. Each update should be version-tagged and assessed for its effect on calibration baselines.

By embracing Calibration Lifecycle Management, organizations gain predictive control over instrument behavior, reduce the risk of unplanned failure, and maintain traceable, auditable calibration records aligned with ISO/IEC 17025 and other quality standards. It's a shift from reactive to proactive metrology governance.

Dual-Method or Cross-Verification

Do not rely solely on one method:

- Validate optical particle counters with flow imaging or SEM.
- Confirm pH with ISE and titration.
- Check flow meters with ultrasonic backups.
- Use redundant or orthogonal technologies where possible.

A key principle in building a resilient measurement system is to never rely solely on a single method or instrument, especially in critical or regulated applications. Measurement errors can result from sensor drift, calibration flaws, matrix effects, or instrument malfunction. Implementing dual-method or cross-verification strategies introduces redundancy and enables early detection of anomalies, ensuring that data is both accurate and trustworthy.

Why It Matters:
Even highly calibrated instruments can deliver incorrect results due to hidden failure modes, such as fouling, optical interference, signal noise,

or software bugs. By comparing results across independent measurement methods, users can detect inconsistencies early before they affect product quality, patient safety, environmental compliance, or financial transactions.

Cross-Verification Strategies:

- Particle Counting:
    - Do not rely exclusively on optical particle counters (OPCs), which are sensitive to refractive index and shape artifacts.
    - Validate results periodically with flow imaging systems or scanning electron microscopy (SEM) for particle morphology and true size verification.
- pH Measurement: Confirm pH values obtained from glass electrodes or ISEs with acid-base titration, especially when working near compliance thresholds or in chemically complex matrices.
- Flow Rate Monitoring: Use ultrasonic or Coriolis flow meters as backup systems for traditional orifice or turbine flow meters, especially in custody transfer or high-value fluid operations.
- Temperature Measurement: Validate infrared thermometer readings with contact thermocouples or blackbody references, especially in clinical or safety screening contexts.
- Elemental Analysis: In metallurgy or pharmaceutical labs, cross-check OES or XRF data with wet chemistry or ICP-OES when discrepancies are observed.

Whenever possible, select redundant or orthogonal technologies—those based on different physical principles—to provide independent confirmation of critical measurements. This layered defense improves confidence in results, strengthens regulatory compliance, and supports rapid root cause analysis when discrepancies arise.

11
Historical Data Normalization Protocol

Abstract
When calibration standards shift, historical data may become misleading. Implement data normalization algorithms to convert pre-change data for compatibility with new standards. Clearly annotate records to distinguish "pre-change" from "post-change" data and establish policies for retrospective data re-evaluation, ensuring critical thresholds and compliance are accurately assessed. Train technicians and engineers to understand calibration processes, recognize drift, and interpret data contextually. One must utilize modern instruments with self-diagnostics to track calibration history, trigger alerts for expired calibrations, and provide exportable logs. These strategies maintain data integrity and ensure reliable, traceable measurements across evolving standards.

Keywords: Experimentation, data integrity, methodology, precision, reproducibility, analysis, discovery

When a standard shifts, older data can become misleading.

Establish:

- Data normalization algorithms
- Clear annotation of "pre-change" vs. "post-change" records
- Policy for retrospective re-evaluation of legacy data

When a reference standard or calibration method changes, it can render historical data misleading or incompatible with current measurements. This poses a serious challenge for organizations relying on long-term trend analysis, quality control baselines, or compliance records. To maintain continuity and data integrity, organizations must implement a Historical Data Normalization Protocol, a structured method for adjusting, annotating, and interpreting past data in light of calibration shifts.

Michael D. Holloway. Accuracy and Calibration Failures in Science and Engineering, (159–164)

Why It Matters:
Without normalization, decisions based on historical trends can become flawed. For example, particle counts before and after the SRM 2806a transition appeared inconsistent, leading to misdiagnosed system cleanliness. Similarly, in medical, environmental, or industrial applications, unadjusted legacy data may mask true performance trends or trigger false alarms.

Key Elements of a Normalization Protocol:

- Data normalization algorithms:
  Develop and apply mathematical models or correction factors to convert "pre-change" data into a form that is comparable to "post-change" results. These algorithms may be derived from side-by-side testing, regression models, or validation studies conducted during the transition.
- Annotation of calibration epochs:
  Clearly label records with metadata that identifies the calibration standard, method version, and measurement conditions. For example, particle count records might indicate "Measured using SRM 2806 (pre-2011)" or "Calibrated per ISO 11171:2016".
- Policy for retrospective data re-evaluation:
  Define when and how legacy data should be retrospectively adjusted or reinterpreted, especially for critical thresholds, regulatory reports, or long-term asset performance assessments. Include this policy in SOPs and quality manuals, and train analysts to recognize the implications of calibration shifts in historical comparisons.

By treating historical data as context-dependent and version-controlled, organizations can preserve the value of past measurements while aligning them with current standards, ensuring accurate benchmarking, compliant reporting, and sound decision-making across time.

Training and Awareness Programs

Technicians, engineers, and lab staff must be trained to:

- Understand how standards are created, maintained, and changed

- Recognize signs of calibration drift
- Interpret data within the context of calibration status

Even the most advanced calibration protocols and systems are only as effective as the people who use them. That's why a critical pillar of any systemic defense in metrology is robust training and awareness programs. These programs ensure that technicians, engineers, lab staff, and quality personnel not only follow procedures but also understand the underlying principles of calibration, traceability, and measurement uncertainty.

Why It Matters:
Many past calibration-related failures, from particle count misinterpretation to thermocouple drift and pressure transducer errors, were not only due to bad equipment but also due to a lack of understanding by end users. When personnel are unaware of how standards evolve or how instrument drift manifests, they are more likely to misinterpret data or overlook early warning signs.

Key Training Objectives:

- Understand how standards are created, maintained, and updated:
  Staff should be familiar with metrological traceability chains, the role of national metrology institutes (e.g., NIST and PTB), and how changes in reference standards (e.g., SRM 2806a) affect real-world measurements. This builds context around why certain procedures exist and why updates must be taken seriously.
- Recognize signs of calibration drift:
  Personnel should be trained to identify subtle indicators of drift, such as unexpected data shifts, inconsistencies between instruments, or measurement instability. This empowers them to take corrective action early before quality or safety is compromised.
- Interpret data within the context of calibration status:
  All measurement results must be viewed through the lens of when and how the instrument was last calibrated. Staff should be able to assess whether data is current and aligned with updated standards and whether normalization or cross-verification is required.

Implementation Tips:

- Integrate training into onboarding, certification, and annual refresher programs.
- Use real-world case studies (e.g., Deepwater Horizon, and CT dosimetry drift) to illustrate the consequences of poor calibration awareness.
- Include calibration awareness in ISO 9001, ISO/IEC 17025, and GMP/GLP compliance training.

With proper training, calibration becomes more than a compliance checkbox; it becomes an embedded culture of measurement excellence, where everyone plays a role in maintaining data integrity.

Instrument Self-Diagnostics and Logs

Modern equipment should:

- Record calibration history
- Trigger alerts when a calibration becomes invalid or expired
- Allow users to export calibration logs for audits

As instrumentation becomes smarter and more integrated into quality systems, self-diagnostics and automated calibration tracking are essential tools for maintaining metrological integrity. Modern equipment, whether in labs, manufacturing, or field applications, should be equipped with features that actively monitor calibration status, alert users to issues, and support auditable traceability.

Why It Matters:
Relying solely on manual tracking of calibration dates or trusting that an instrument is "still fine" leads to blind spots. Drift can occur silently, and expired calibrations can go unnoticed, especially in busy operations or decentralized systems. Built-in diagnostics ensure that calibration control is continuous, not reactive.

Key Capabilities of Modern Instruments:

- Calibration history recording:
  Devices should automatically log every calibration event, including the date, standard used, technician ID (if applicable), and method or software version. These logs form a digital chain of traceability, critical for internal QA and external audits.
- Real-time calibration alerts:
  Instruments must trigger alerts or lockouts when calibration is due, expired, or when a significant drift or fault is detected during self-check routines. This prevents uncalibrated use and supports proactive maintenance scheduling.
- Exportable calibration logs:
  Users should be able to easily export calibration histories and diagnostics data in standard formats (CSV, PDF, and XML) for inclusion in audit reports, CAPA records, or regulatory submissions. Some systems also offer cloud integration or API access for seamless LIMS/QMS integration.

Advanced Features (where applicable):

- Built-in calibration references or self-check standards for instruments like balances, spectrophotometers, flow meters, and analyzers.
- Audit trails with timestamps and user authentication for regulated environments (e.g., 21 CFR Part 11 compliance).
- Drift trend visualization to support predictive maintenance and instrument replacement planning.

By embedding calibration intelligence directly into the instrument, organizations can reduce human error, enforce compliance, and maintain a continuous, transparent record of measurement validity, ensuring that data remains trustworthy, traceable, and audit-ready at all times.

12

External Audits and Peer Comparisons

Abstract
A robust calibration system requires external validation through peer comparison and third-party audits. Round-robin testing, blind sample evaluations, and inter-site comparisons reveal hidden biases and inconsistencies in measurements. Industry-wide data repositories provide a centralized database for calibration issues, standard updates, and error analytics, promoting cross-industry learning. An ethical culture of measurement integrity emphasizes uncertainty disclosure, transparent documentation of calibration failures, and recognition of instrument fallibility. These practices enhance measurement reliability, support compliance, and foster continuous improvement by embedding transparency, accountability, and collaborative learning into metrology systems.

***Keywords*:** Pollution control, ecosystem health, sampling, detection, monitoring, reporting, sustainability

Participate in:

- Round-robin testing/proficiency programs
- Blind sample evaluations
- Inter-lab or inter-site comparisons to detect outliers

A strong calibration system doesn't operate in isolation; it benefits greatly from external validation and benchmarking. Participating in peer comparison programs and third-party audits adds an essential layer of oversight, helping to detect hidden inconsistencies, uncover bias, and validate the accuracy and repeatability of measurements across different sites, labs, or instruments.

Why It Matters:
Internal calibration systems can sometimes drift into systemic bias without being detected, especially when all measurements are self-referenced. By engaging with external bodies and peer institutions, organizations gain

Michael D. Holloway. Accuracy and Calibration Failures in Science and Engineering, (165–170)

objective confirmation of their measurement integrity and can identify outlier behavior before it leads to quality issues or compliance failures.

Key Practices:

- Round-robin testing (inter-laboratory comparisons): Multiple labs measure the same standardized sample, and results are statistically compared. This process reveals deviations in technique, instrument performance, or calibration accuracy. It's especially important in metals testing, environmental analysis, and pharmaceutical QC.
- Blind sample evaluations: Labs receive unknown reference samples and must report results without prior knowledge of the expected values. This tests real-world performance and helps assess analytical accuracy under standard operating conditions.
- Inter-site comparisons: For organizations with multiple facilities or global operations, regular inter-site calibration comparisons ensure consistent measurement practices across the enterprise. Discrepancies can point to training gaps, environmental effects, or equipment differences.

Benefits:

- Identifies systematic errors and methodological bias that internal checks might miss.
- Supports ISO/IEC 17025, ISO 9001, and GMP/GLP compliance.
- Builds credibility with clients, regulators, and partners through external verification of measurement quality.
- Encourages a culture of transparency, continuous improvement, and collaboration.

By integrating external audits and peer comparisons into your calibration system, you gain more than just benchmarking; you gain a clear, objective lens on your measurement reliability and a roadmap for ongoing improvement.

Industry-Wide Data Repositories and Alerts

Like a "metrology equivalent" of an NTSB (National Transportation Safety Board):

- Centralized database of known calibration issues or shifts
- Notification system for when standards are updated
- Industry reports on common sources of measurement error

To achieve true systemic resilience in measurement systems, there must be an industry-level infrastructure for sharing calibration-related knowledge, alerts, and lessons learned. Much like the NTSB (National Transportation Safety Board) in aviation and transportation, the metrology field would benefit greatly from a centralized, transparent platform that collects, analyzes, and distributes information about calibration issues, standard changes, and recurring measurement challenges.

Why It Matters:
Most organizations discover calibration drift or standard transitions independently, often after quality failures, compliance breaches, or safety incidents. Without shared visibility, entire industries repeat the same mistakes, wasting resources and risking harm. A collective repository would help prevent recurrence, accelerate awareness, and standardize responses to measurement-related risks.

Key Features of a Metrology Knowledge Hub:

- Centralized database of calibration issues and alerts:
  A structured, searchable repository that documents known cases of calibration drift, failed instruments, reference standard transitions, and software-induced measurement errors across industries (e.g., SRM 2806a transition, FID drift, and ISE degradation).
- Standard update notifications:
  A subscription-based alert system for new or revised standards from NIST, ISO, ASTM, IEC, and others. Stakeholders, including labs, OEMs, auditors, and maintenance teams, would be automatically notified of changes affecting their operations.

- Root cause and incident reports:
  Industry-shared reports on measurement failures, out-of-tolerance events, and recall triggers, similar to how the NTSB publishes investigation findings. These would include technical causes, consequences, and mitigation strategies.
- Common error analytics and trend tracking:
  Aggregated data showing frequent sources of measurement error across sectors (e.g., most common instruments showing drift, and seasonal effects on certain sensors), guiding equipment upgrades, and training priorities.

Benefits:

- Promotes early detection and cross-industry learning, avoiding repeat incidents.
- Supports standardized protocols for measurement change management.
- Informs training programs, QA systems, and procurement decisions with real-world calibration risk data.
- Fosters a collaborative metrology ecosystem, where knowledge improves reliability for all.

Creating a "metrology equivalent of the NTSB" would elevate measurement quality from a localized practice to an industry-wide safety net, embedding shared intelligence into the foundation of technical and regulatory trust.

Ethical Culture of Measurement Integrity

Promote an organizational mindset that values:

- Measurement uncertainty disclosure
- Honest documentation of calibration failures
- Recognition that even "good" instruments can go bad silently

At the heart of any resilient metrological system is not just the equipment or procedures but the mindset of the people using them. Fostering an ethical culture of measurement integrity means building an environment where truth in measurement is non-negotiable and where transparency, humility, and accountability are woven into daily practice. It is a culture that values precision not just in instruments, but in judgment and reporting.

This concept is especially important in high-stakes industries like pharmaceuticals, aerospace, energy, and environmental science, where measurement errors can lead to product recalls, regulatory penalties, safety incidents, or worse.

3 Key Principles of an Ethical Measurement Culture

1. Measurement Uncertainty Disclosure
    - What It Means: Always communicate the known uncertainty associated with any measurement. No reading is absolute; every result carries a confidence interval or tolerance band.
    - Why It Matters: Presenting results without uncertainties gives a false sense of precision and may lead to overconfidence in borderline decisions (e.g., release testing, regulatory limits, and environmental compliance). Disclosing uncertainty is an act of scientific honesty and is critical for informed risk management.
    - Best Practices:
        - Include uncertainty estimates in reports, control charts, and dashboards.
        - Train staff to interpret and communicate uncertainty, not just final values.
        - Use control limits or guard bands that account for measurement uncertainty, not just nominal specifications.
2. Honest Documentation of Calibration Failures
    - What It Means: When an instrument fails calibration, it must be openly documented, investigated, and communicated—no hiding, no soft-pedaling, and no quietly adjusting tolerances to pass.
    - Why It Matters: Ignoring or minimizing failed calibrations undermines data traceability, quality assurance, and regulatory compliance. More importantly, it risks making decisions based on invalid data, which can jeopardize safety, reliability, and public trust.

- Best Practices: Treat failed calibrations as nonconformances requiring formal investigation (CAPA if necessary).
  - Notify stakeholders about potentially affected data and initiate re-analysis where required.
  - Preserve audit trails of failures and corrections. Use them to drive process improvement rather than conceal issues.

3. Recognition That Even "Good" Instruments Can Go Bad Silently
   - What It Means: Acknowledge that instruments are not infallible; even the most reliable device can drift, degrade, or misbehave without warning.
   - Why It Matters: Many measurement failures stem not from poor instruments, but from overreliance on past performance. Assuming "it's fine because it's always been fine" is a mindset that creates blind spots.
   - Best Practices:
     - Promote a culture of questioning and verification, not blind trust.
     - Encourage routine cross-checks, dual-method validations, and inter-instrument comparisons.
     - Use data analytics and drift logs to detect subtle trends before they become out-of-tolerance failures.
     - Empower technicians to flag anomalies without fear of blame; treat suspicion as a prompt for action, not dismissal.

13

How to Build and Sustain This Culture

Abstract

Leadership commitment drives a culture of measurement integrity. Integrate transparency and curiosity into management practices. Train staff on measurement ethics and traceability. Reward accuracy over convenience. Embed ethical principles in SOPs and calibration policies. Facilitate storytelling and knowledge sharing through forums and newsletters. Establish a comprehensive calibration management program with metrological traceability, secure calibration records, and change management controls. Implement redundancy and cross-verification. Preserve historical data integrity with annotations and normalization. Ensure continuous training and competency. Utilize instrument self-monitoring features. Participate in external validation and audits. Maintain standard awareness and promote a culture of transparency and accountability.

***Keywords*:** Risk management, hazard prevention, system integrity, maintenance, safety protocols, reliability engineering, incident analysis

- Leadership Commitment: Set the tone from the top. Quality leaders and technical managers must model transparency, curiosity, and a no-blame approach to measurement issues.
- Training and Awareness: Include ethics of measurement, uncertainty, and traceability in onboarding and recurring training. Share real-world case studies where measurement failures had consequences.
- Reward Integrity Over Convenience: Recognize and reward personnel who catch potential errors, challenge assumptions, or delay a process to ensure measurement quality. Reinforce that accuracy and honesty trump speed and smoothness.
- Quality System Integration: Embed ethical metrology principles into SOPs, calibration policies, and deviation handling. Ensure procedures do not only ask, "Did it pass?" but also, "Does it still make sense?"

Michael D. Holloway. Accuracy and Calibration Failures in Science and Engineering, (171–178)

- Storytelling and Knowledge Sharing: Create forums, newsletters, lunch-and-learns, and case briefings where lessons from measurement incidents (internal or industry-wide) are shared. Make measurement integrity part of the organization's story.

A culture of measurement integrity is not built overnight. It takes deliberate effort, leadership buy-in, and a willingness to embrace uncertainty as a strength, not a weakness. By promoting ethical practices in how data is measured, reported, and acted upon, organizations can move from compliance to excellence and from trust to truth.

> "Trustworthy measurement systems are not built by better instruments alone but by better systems, better people, and better governance."

Establishing a Comprehensive Calibration Management & Measurement Integrity Program

Elements of the Standard Operating Procedure

Purpose:
To define the procedures, responsibilities, and controls necessary to ensure all measurements are traceable, reliable, and ethically managed through lifecycle-based calibration, change management, verification, and competency practices.

1. Establish Metrological Traceability
   Objective: Ensure all measurements can be traced to recognized national/international standards (e.g., SI units).
   - Use only calibration services accredited to ISO/IEC 17025.
   - Maintain certificates of calibration referencing SI-traceable standards (e.g., NIST and PTB).
   - Require uncertainty budgets and traceability chains in all calibration documentation.
   - Link each measurement instrument to its current calibration reference in the asset registry.
2. Maintain Calibration Records & Logs
   Objective: Build a secure, traceable log of all calibration activities.

- Create a centralized calibration management system or database.
- Log:
  - Calibration dates and due dates
  - Instrument ID, model, and serial number
  - Calibration methods, standards, and personnel
  - Pre- and post-calibration values
  - Any adjustments or failures
- Ensure tamper-proof records with audit trails for regulated environments.

3. Implement Change Management Controls
   Objective: Ensure calibration-related changes are managed, communicated, and documented properly.
   - Establish a Standard Change Notification (SCN) process.
     - Triggered by standard updates, software changes, method shifts, or calibration vendor switches.
   - Include:
     - Stakeholder notifications
     - Impact analysis (technical, data continuity, regulatory)
     - Guidance for comparing pre- and post-change data
     - Versioning/timestamping of calibration results
4. Enable Redundancy & Cross-Verification
   Objective: Detect and prevent measurement errors *via* independent verification.
   - Use dual-method validation where feasible:
     - OPCs validated by SEM or flow imaging
     - pH by titration and electrode
     - Flow meters *via* ultrasonic/Coriolis backups
   - Perform routine cross-instrument checks to detect silent drift.
   - For critical systems, deploy redundant measurement paths.
5. Preserve Historical Data Integrity
   Objective: Maintain usable, contextualized measurement records across calibration transitions.
   - Annotate measurement records with:
     - Calibration version, standard reference (e.g., SRM 2806a)
     - Instrument firmware/software version
   - Develop normalization algorithms for comparing pre/post-calibration-shift data.

- Establish a protocol for retrospective review of legacy data if new standards are adopted.

6. Build a Training & Competency Program
   Objective: Ensure personnel understand and apply calibration principles correctly.
   - Provide recurring training on:
     - Calibration theory and traceability
     - Uncertainty, measurement ethics, and data interpretation
     - Change management and SOP compliance
   - Certify competency levels based on roles (technician, QA lead, lab manager).
   - Maintain training records and schedule periodic refreshers.
7. Leverage Instrument Self-Monitoring Features
   Objective: Automate early warning and performance validation.
   - Use instruments with:
     - Internal diagnostics and calibration reminders
     - Drift detection and auto-verification
     - Firmware version logs and timestamps
   - Configure alerts for:
     - Expired calibration
     - Self-test failures
     - Abnormal drift or out-of-spec readings
8. Participate in External Validation & Audits
   Objective: Benchmark accuracy and detect systemic bias.
   - Join:
     - Round-robin (inter-lab) comparisons
     - Blind proficiency tests
     - External QA audits (regulatory or independent)
   - Use discrepancies to improve methods, retrain staff, or recalibrate instruments.
9. Ensure Standard Awareness & Communication
   Objective: Keep the organization aligned with evolving calibration standards.
   - Subscribe to standard bodies (NIST, ASTM, ISO) for alerts.
   - Maintain an internal standards registry with summaries of key changes.

- Issue SCN bulletins and SOP updates upon standard revisions.
- Assign a metrology change coordinator to track and disseminate updates.

10. Promote a Culture of Measurement Ethics

    Objective: Embed transparency, humility, and accountability in measurement practices.
    - Train staff to:
      - Report calibration anomalies or failures without fear
      - Disclose uncertainty in reports
      - Understand that instruments can "fail silently"
    - Reward:
      - Integrity in reporting and stopping processes when measurement trust is compromised
      - Root cause investigations that prevent future errors
    - Incorporate measurement ethics into quality policy and leadership messaging.

    Documentation & Audit Readiness
    - Maintain SOPs, training records, SCNs, calibration certificates, change logs, and audit findings in a validated document control system.
    - Prepare evidence trails for:
      - Internal QA review
      - External audits (FDA, ISO, EPA, customer inspections)

    Optional Enhancements
    - Integrate calibration systems with LIMS or ERP platforms.
    - Use AI/ML tools to predict instrument failure or calibration drift.
    - Publish internal "lessons learned" briefs from incidents or audits.

Calibration Audit Checklist

Audit category	**Checklist item**	**Evidence to review**
Metrological traceability	Are all instruments traceably calibrated to SI units through an accredited lab?	Calibration certificates with traceability chain

(*Continued*)

(*Continued*)

Audit category	Checklist item	Evidence to review
Calibration records & logs	Is each instrument's calibration log complete, timestamped, and versioned?	Instrument history logs, software versions, calibration reports
Change management	Is there a formal process to assess and communicate the impact of calibration standard changes?	Standard Change Notifications, meeting minutes, memos
Redundancy & cross-verification	Are critical measurements verified by an independent or orthogonal method periodically?	Dual-method test reports, cross-check logs, exception reports
Historical data integrity	Is historical data annotated or normalized following calibration standard shifts?	Data system annotations, recalibration reports, normalization protocols
Training & competency	Are all relevant staff trained in interpreting calibrated data and understanding metrological changes?	Training records, competency assessments, refresher logs
Instrument self-monitoring	Do instruments have self-diagnostics, drift detection, or calibration alerts enabled?	Instrument configuration files, screenshots of alerts/ logs
External validation & audits	Does the organization participate in inter-lab comparisons or round-robin testing?	External audit reports, round-robin results, third-party reviews

(*Continued*)

(*Continued*)

Audit category	Checklist item	Evidence to review
Standard awareness & communication	Are staff notified when relevant standards (e.g., ISO, ASTM, and SRM) are revised or replaced?	Email alerts, policy documents, standards library access
Measurement ethics & culture	Is there a documented culture of transparency and integrity regarding measurement uncertainty and calibration status?	Internal quality policies, incident logs, whistleblower protections

14
Steps for Suggested Safeguards

Abstract

Embedding safeguards in complex systems design and operation enhances resilience against failures. Implement redundancy and challenge loops through independent reviews and red teams to uncover blind spots. Ensure data transparency by documenting limitations and uncertainty in measurements. Foster cross-functional decision-making with protocols that translate technical insights into actionable executive understanding. Formalize change detection systems to trigger alerts and impact assessments when standards or algorithms change. Design interfaces that highlight anomalies and train users to recognize and question discrepancies. Institutionalize learning through postmortems, and stress-test systems with scenario simulations to prepare for worst-case conditions.

Keywords: Production efficiency, process control, equipment calibration, quality management, automation, standardization, yield improvement

This chapter focused on establishing safeguards through a series of easy-to-follow steps. To prevent future high-impact failures across technical and organizational domains, a proactive framework of safeguards must be embedded into the design, operation, and oversight of complex systems. This includes building redundancy and challenge loops such as independent reviews and red teams to expose blind spots and failure modes early. Data transparency and contextual integrity are critical, requiring clear communication of limitations, uncertainties, and assumptions in all critical measurements and forecasts. Decision-making must be cross-functional, with structured protocols to translate technical insights into actionable executive understanding. Systems should incorporate formalized change detection mechanisms that automatically trigger stakeholder alerts, impact analyses, and updates to data interpretation when standards, algorithms, or inputs change. Interfaces must be designed with human factors in mind, emphasizing clarity, anomaly detection, and training users not only to operate systems but to also recognize and question when something is wrong. Organizations should also institutionalize retrospective learning through

Michael D. Holloway. Accuracy and Calibration Failures in Science and Engineering, (179–198)

blameless postmortems and lessons-learned registries. A resilient culture embraces healthy skepticism and inquiry, encouraging whistleblowers and protecting dissenting voices. Finally, resilience must be stress-tested regularly through scenario simulations and worst-case drills, ensuring systems hold up not just in normal conditions but at the edges of design tolerance. Together, these safeguards form a multilayered defense against complacency, model overreliance, and decision-making in the dark.

Build Systems with Redundancy and Challenge Loops

- Require independent review of critical models, data, and decisions.
- Encourage "devil's advocate" roles or red teams to explore failure modes.

Building systems with redundancy and challenge loops means designing not only for performance and efficiency but also for resilience and error detection. In complex environments where the consequences of failure can be catastrophic, this approach acts as a systemic immune response, catching issues that slip past standard workflows. It emphasizes independent validation, cognitive diversity, and the intentional cultivation of dissent to uncover blind spots, flawed assumptions, and hidden failure modes.

Definition of Redundancy and Challenge Loops

- Redundancy: Incorporating independent, parallel systems, models, or personnel to validate outputs or decisions. Redundancy is not just about backup systems; it is about parallel interpretation, assessment, or execution to catch errors before they propagate.
- Challenge Loops: Embedding structured critique and questioning mechanisms into the decision-making process, such as red teaming, devil's advocate roles, or contrarian reviews, to deliberately test and interrogate critical assumptions, data interpretations, and model boundaries.

Step-by-Step Process to Build Redundancy and Challenge Loops
Step 1: Identify Critical Decision and Risk Points

- Map out where decisions are made based on models, sensor data, or human judgment.
- Highlight points with high consequences if errors go undetected (e.g., launch go/no-go decisions, patient treatment protocols, and asset management triggers).
- Determine what assumptions or models are being relied upon at those points.

Step 2: Define the Redundancy Requirements

- Establish independent parallel systems where possible:
  - Dual-model validation (e.g., two independently developed forecasting models).
  - Redundant sensors measuring the same parameter using different physical principles (e.g., ultrasonic and Coriolis flow meters).
  - Independent analysts reviewing critical data pipelines.
- Ensure these redundant systems:
  - Operate on different algorithms, data sources, or assumptions to avoid "shared blind spots."
  - Report discrepancies instead of averaging them out.

Step 3: Formalize Independent Review Protocols

- Create standing procedures for:
  - Peer review of all critical models and assumptions before deployment.
  - Regular audit cycles where independent teams review recent decisions and their underlying data.
- Define minimum review thresholds, such as:
  - Any model influencing financial, safety, or compliance decisions must be independently validated before use.
  - Data models updated post-deployment must be re-reviewed after significant changes.

Step 4: Establish Challenge Loops and Red Teams

- Designate "devil's advocate" roles in planning or review meetings. Their job: to find what could go wrong.
- Build a Red Team program:
  - Form a cross-disciplinary team with no stake in the project outcome.
  - Assign them to explore how the system might fail technically, operationally, or socially.
  - Empower them to conduct stress tests, simulate anomalies, or intentionally misinterpret assumptions to test system resilience.
- Use tools like pre-mortem analysis, where the team assumes a failure has occurred and works backward to find what went wrong.

Step 5: Compare, Document, and Resolve Divergences

- Use structured processes to compare outputs from primary and redundant systems.
- When discrepancies occur:
  - Flag for escalation.
  - Log divergences and force a "discrepancy resolution meeting" before proceeding.
  - Document decision rationale if moving forward despite unresolved contradictions.

Step 6: Integrate into Governance and Workflow

- Build review cycles into SOPs, QA plans, and safety checklists.
- Require challenge loop sign-off for mission-critical operations (e.g., code deployment, clinical decisions, and crisis responses).
- Align with existing risk management frameworks and ensure these loops are auditable.

Step 7: Evaluate and Evolve the Loop

- After major projects or anomalies:

- Evaluate how well the redundancy and challenge mechanisms worked.
- Update checklists, review protocols, and team structures based on findings.
- Conduct lessons-learned workshops with red teams and primary teams together to foster shared insight.

Cultural Considerations

To truly succeed, challenge loops require a supportive organizational culture:

- Promote intellectual humility, the willingness to be wrong.
- Reward those who catch problems, not just those who deliver results.
- Protect and value dissent and constructive criticism, especially from junior staff or outsiders.
- Ensure red teams and challengers have access to data, tools, and decision-makers, not just a seat at the table

Feature	Benefit
Redundant Systems	Catch technical or model-specific failures
Independent Review	Break echo chambers and confirm assumptions
Red Teaming	Discover novel or unexpected failure modes
Challenge Loops	Foster better reasoning and decision-making under pressure
Documentation of Discrepancies	Build institutional memory and traceability
Cultural Reinforcement	Normalize questioning and transparency

In essence, redundancy and challenge loops are not inefficiencies; they are strategic safeguards that protect high-stakes decisions from the fragility of human assumption and systemic bias. In a world increasingly driven by complex models, automation, and rapid decisions, they are no longer optional; they are essential.

Improve Data Transparency and Context Awareness

- Ensure source data is always qualified; document limitations, assumptions, and confidence levels.
- Use uncertainty reporting in all critical measurements and forecasts.

From global policy to machine learning models to real-time industrial control, transparency and contextual understanding of data are no longer optional. They are essential pillars of integrity, trust, and performance. Without clearly defined limits, assumptions, and confidence levels, even the most precise datasets can mislead users, creating a dangerous illusion of certainty. By embedding data transparency and context awareness into every layer of the data lifecycle, organizations can ensure that decisions are informed, not just data-driven.

What Do We Mean by Data Transparency and Context Awareness?

- Data transparency means that all users, analysts, decision-makers, auditors, and regulators can clearly see how data was generated, what it represents, what its quality is, and how it should be used.
- Context awareness ensures that data is accompanied by metadata, assumptions, methods, and uncertainty ranges so that it is interpreted correctly and not used beyond its intended scope.

Step-by-Step Process to Improve Data Transparency and Context Awareness

Step 1: Define and Tag Source Data Clearly

- For every dataset, sensor feed, or model input:
  - Record the source (device, system, vendor, manual entry).
  - Document data type, unit of measure, and timestamp format.
  - Include data lineage: who collected it, when, and under what conditions.
- Use metadata tagging systems to store this information alongside the data.

Step 2: Document Assumptions and Limitations

- Create a data declaration sheet for critical datasets and models that includes:
  - Known limitations (e.g., sensor drift, sample bias, and historical gaps).
  - Assumptions behind data collection (e.g., uniform population behavior and standard lab temperature).
  - Scope of validity (e.g., "only valid for temperatures between -20°C and 50°C").
- Maintain these declarations in version-controlled documentation (e.g., embedded in dashboards, wikis, or data catalogs).

Step 3: Integrate Uncertainty Reporting

- For all critical measurements, forecasts, or model outputs, require:
  - Confidence intervals or standard deviations.
  - Probability ranges or percentiles (e.g., "90% chance that revenue will fall between $2.1M and $2.8M").
  - When applicable, Bayesian updates to show how confidence evolves with new data.
- Display uncertainty visually in tools and dashboards (e.g., shaded bands, error bars, and confidence zones).

Step 4: Enforce Context-Rich Data Presentation

- In reports and visualizations:
  - Require source attribution and method summary.
  - Show when the data was last updated and under what calibration or software version.
  - Provide links to data provenance documentation and assumptions.
- Avoid charts or tables that present absolute numbers without qualifiers; replace them with "qualified data narratives".

Step 5: Build Feedback Mechanisms

- Allow users to flag questionable data, outdated assumptions, or inconsistencies.

- Periodically conduct "data trust audits" to test:
    - Whether users understand the meaning and boundaries of the data.
    - How well confidence levels and uncertainty are being respected in decisions.

Step 6: Include Transparency in Data Governance Policies

- Make data transparency and uncertainty disclosure formal policy requirements.
- Embed them into:
    - Data acquisition checklists.
    - Model development protocols.
    - Regulatory compliance documentation (e.g., ISO 9001, ISO/IEC 17025, and FDA Part 11).

Step 7: Train Staff to Interpret Data in Context

- Offer regular workshops or modules on:
    - Understanding and applying confidence intervals and error margins.
    - Recognizing when data may be incomplete or misapplied.
    - Communicating uncertainty effectively to non-technical stakeholders.
- Encourage a shift from "What does the data say?" to "What can this data support, and with what level of confidence?"

Example in Practice
Poor Practice:
A dashboard shows, "Machine vibration = 3.6 mm/s—within normal range." No mention of sensor type, calibration age, uncertainty, or environmental conditions.

Improved Practice:
"Machine vibration = 3.6 ± 0.4 mm/s (measured *via* Type X sensor, calibrated March 2024). Confidence: 95%. Last reading captured at 28°C. Valid for ambient temperature range 10°C to 40°C."

Practice	Benefit
Documenting assumptions	Prevents misuse or overgeneralization of data
Uncertainty reporting	Supports risk-informed decision-making
Metadata tagging	Improves traceability and auditability
Visual context indicators	Enhances interpretation and user trust
Training for context awareness	Builds analytical maturity across the organization

Why It Matters

- Data without context is like medicine without dosage—potentially harmful if misunderstood.
- Overconfidence in "clean" numbers without error bars can lead to bad investments, missed warning signs, or misinformed policy.
- Transparency is not just a best practice; it is a defense mechanism against overfitting, bias, and misjudgment.

Improving data transparency and context awareness is about treating data as a dialog, not a dictate. It ensures that users not only know what the numbers are but also what they mean, where they come from, and how much they can be trusted. In our world, decisions are increasingly data-driven; context is the difference between clarity and catastrophe.

Enforce Cross-Functional Decision-Making

- Bridge technical, operational, and executive levels with common understanding.
- Create translation protocols for converting technical findings into executive decisions.

Siloed thinking and fragmented communication are not just inefficiencies; they are risks. Cross-functional decision-making ensures that technical insights, operational realities, and executive imperatives are aligned in both understanding and action. When decisions are made without bridging these domains, organizations risk making technically sound but operationally unfeasible or strategically expedient but technically flawed choices.

To avoid this, organizations must intentionally structure collaboration and communication so that decisions benefit from diverse perspectives and shared accountability. This means developing systems that can translate technical findings into executive insight while also ensuring that business goals and operational constraints inform technical direction.

Step-by-Step Process to Enforce Cross-Functional Decision-Making
Step 1: Identify Key Decision Domains and Stakeholders

- Map out critical decision points where technical, operational, and strategic input is required (e.g., system commissioning, risk acceptance, incident response, and product release).
- Define stakeholder groups:
  - Technical leads (engineers, data scientists, analysts)
  - Operational leads (maintenance, production, safety, logistics)
  - Executives (finance, compliance, strategy)

Step 2: Establish Cross-Functional Review Forums

- Create recurring joint review boards or decision councils where all three layers—technical, operational, and executive—are represented.
- Use these boards for:
  - Project milestone approvals
  - Major risk acceptance decisions
  - Incident reviews and postmortems
- Assign a decision integrator or facilitator responsible for aligning the language and priorities across roles.

Step 3: Build Translation Protocols for Technical Communication

- Require technical teams to include the following in reports and presentations:
  - Plain-language summaries of findings and implications.
  - Assumptions, limitations, and uncertainty range of results.

  - Operational relevance: how findings affect workflow, safety, or uptime.
  - Strategic impact: what the data means for KPIs, customer experience, cost, or compliance.
- Use visual aids, analogies, and risk heat maps to bridge expertise gaps.

Step 4: Standardize Cross-Disciplinary Documentation Templates

- Develop decision memos that include:
  - Technical diagnosis or forecast
  - Operational impact assessment
  - Strategic implications
  - Recommended action plan with risk trade-offs
- Require that all critical recommendations be accompanied by cross-sign-off (e.g., signed by the technical lead, operations, and exec sponsor).

Step 5: Enable Two-Way Knowledge Exchange

- Provide executives with basic technical literacy training relevant to the systems they govern (e.g., data interpretation, model limitations, and safety systems).
- Provide technical and operational teams with business acumen training to help them understand financial and strategic considerations.
- Foster mentoring or shadowing programs between departments to build empathy and shared context.

Step 6: Implement Escalation and Override Protocols

- Define conditions under which technical staff can escalate concerns to executives, bypassing middle layers if urgent risks are identified.
- Create documented override procedures for when business needs conflict with technical warnings, requiring:
  - Justification
  - Documented risk acceptance
  - Contingency plans

Step 7: Monitor and Audit Decision Quality

- After major decisions or failures:
  - Review whether all relevant voices were present and heard.
  - Assess if communication gaps contributed to flawed outcomes.
- Include cross-functional representation in postmortem reviews, and identify opportunities to improve integration for future cases.

Cultural and Organizational Enablers

- Flatten the perceived hierarchy in cross-functional meetings; make it safe for technical voices to speak candidly.
- Promote shared accountability; decisions are owned collectively, not blamed individually.
- Reward team-based problem-solving, not just individual performance.

Feature	**Benefit**
Cross-functional boards	Integrated decision-making and accountability
Translation protocols	Better understanding across technical and executive domains
Shared documentation	Clearer alignment of technical facts and business impacts
Bidirectional training	Increases mutual respect and operational fluency
Escalation paths	Prevents suppression of critical information

In Practice: A Brief Scenario

A data scientist develops a predictive maintenance algorithm that flags early bearing wear. The model shows promise but includes a 12% false positive rate. Operations are worried about wasted downtime, and executives are unsure how to weigh the cost of premature intervention. Through cross-functional review:

- Technical staff explain the model's structure, confidence levels, and possible tuning options.
- Operations quantifies the cost of false positives versus an unplanned failure.
- Executives evaluate strategic risk (safety, reputation) and budget impact.

Together, they reach a decision that accounts for all dimensions and documents the rationale for traceability and future learning.

Enforcing cross-functional decision-making is not about meetings or checkboxes; it is about ensuring that no critical decision is made in a vacuum. It is about creating a shared language and structure that allows deeply technical insights to influence strategic direction, and vice versa. In today's interconnected world, the smartest decision is rarely made by the smartest person; it is made by the best-informed team.

Formalizing Change Detection & Notification Systems

- Any change to calibration, standards, data sources, or algorithms should trigger:
- Impact assessment
- Stakeholder notification
- Update of legacy data interpretations

In any data-driven or precision-critical environment, change is inevitable, whether through calibration updates, algorithm revisions, new standards, or shifts in data sources. However, undocumented or poorly communicated changes can silently corrupt trends, invalidate historical comparisons, and lead to misinterpretation of critical measurements or forecasts. That is why organizations must formalize change detection and notification systems: structured processes that track when changes occur, assess their impact, notify stakeholders, and update legacy data or procedures accordingly.

Such systems act as "version control for decision-making," providing a traceable, auditable path for how and why data and the systems that produce it evolve over time. Without them, seemingly minor updates can have cascading effects, from compliance violations and technical errors to lost stakeholder trust.

Step-by-Step Process to Formalize Change Detection & Notification Systems

Step 1: Define What Qualifies as a Critical Change

- Create a formal change classification schema to define what types of changes must trigger the system. Examples:
  - Calibration changes (instrument recalibration, new standards like SRM 2806a)
  - Algorithm/model updates (e.g., changes in coefficients, logic, or feature sets)
  - Data source changes (sensor swap, schema updates, vendor switch)
  - Standard or regulatory changes (ISO revisions, EPA/ASTM changes)
  - Firmware/software upgrades that affect measurement interpretation

Step 2: Deploy Change Detection Mechanisms

- Implement automated change logging systems:
  - For digital systems: Use Git or version control tools to track algorithm and data model updates.
  - For instruments: Enable internal audit trails that log calibration dates, firmware versions, and settings changes.
- Configure automated alerts to flag when:
  - A new calibration certificate is issued.
  - Software or firmware is updated.
  - Data schemas or source integrity checks change.
- Combine system-driven detection with manual reporting protocols for changes initiated by technicians, analysts, or vendors.

Step 3: Conduct a Structured Impact Assessment

For every detected change:

- Assign a change analyst or cross-functional review team.
- Assess:
  - Scope of impact (e.g., which data streams, departments, reports, or compliance standards are affected?)
  - Magnitude of impact (Are historical thresholds or KPIs invalidated?)

    - Whether legacy data remains valid or needs reinterpretation
- Document:
    - Change description
    - Source and reason for change
    - Expected effects on current and historical operations
    - Proposed mitigation steps (e.g., re-benchmarking and normalization)

Step 4: Notify Stakeholders

- Based on the impact assessment, notify all relevant parties, including the following:
    - Engineers/technicians who rely on the affected data
    - Data analysts and scientists working with historical comparisons
    - Compliance officers or QA teams responsible for traceability
    - Executives and decision-makers depending on trend-based KPIs
- Use multi-channel delivery:
    - Internal alert systems (ticketing, dashboards)
    - Email summaries with visual change logs
    - Documentation updates on intranet or shared knowledge bases

Step 5: Update Legacy Data Interpretation

- Where applicable:
    - Annotate or flag historical data as "pre-change" vs. "post-change" in databases and reports.
    - Apply normalization or conversion factors to make older data comparable with new outputs.
    - Update SOPs, dashboards, and models that reference historical baselines.
- Ensure all revisions are:
    - Timestamped
    - Version-controlled
    - Linked to the change log entry

Step 6: Maintain a Centralized Change Register

- Establish a living "Change Register" that logs all changes, assessments, and notifications:
  - Date and type of change
  - Responsible parties
  - Affected systems
  - Links to documentation, data patches, or recalibration files
  - Status: Reviewed/Implemented/Archived
- Review this log regularly in internal audits, QA meetings, or regulatory inspections.

Step 7: Train Staff on Change Awareness

- Train data handlers, system users, and decision-makers to:
  - Recognize when a change has likely occurred
  - Access the change log or change declaration documents
  - Know how to report undocumented changes
- Incorporate change literacy into onboarding and annual compliance training.

Cultural and Organizational Enablers

- Promote accountability without blame; encourage proactive reporting of changes without fear of reprimand.
- Treat every change as an opportunity to increase system transparency, not a nuisance to suppress.
- Align change detection processes with existing quality frameworks like ISO 9001, ISO/IEC 17025, or GxP standards.

Feature	**Benefit**
Change classification schema	Prevents unnoticed drift in calibration, logic, or data
Structured impact assessment	Informs decisions about reprocessing, reinterpretation
Stakeholder alerts	Keeps everyone aligned and reduces surprise failures

(*Continued*)

(*Continued*)

Feature	Benefit
Change register	Ensures traceability and regulatory readiness
Legacy data updates	Preserves continuity and analytical comparability

Formalizing change detection and notification systems creates a resilient backbone for data governance and operational integrity. In complex systems, small, unseen changes can accumulate into major inconsistencies or blind spots. By building change awareness into your infrastructure from code to calibration, you transform change from a lurking liability into a managed, visible, and actionable process. This is how reliable organizations learn from the past, adapt to the present, and prepare for the future.

Prioritize Human Factors in Interface Design

- Design interfaces that clearly indicate the system state and highlight anomalies.
- Train users not just on operation, but on failure recognition and data skepticism.

We are increasing automation and complex system interdependencies; the interface is often the last and sometimes the only link between critical data and human decision-making. When systems fail, it is often not because the data was absent, but because it was misunderstood, misrepresented, or missed entirely. This makes human factors in interface design not just a matter of usability, but of safety, reliability, and operational resilience.

Prioritizing human factors means designing interfaces that are intuitively navigable, contextually informative, and cognitively aligned with how users perceive, process, and act on information, especially under pressure. It also means training users to question what the system shows, not just to accept it.

Step-by-Step Process to Prioritize Human Factors in Interface Design
Step 1: Understand the User's Mental Model and Work Context

- Conduct user research and task analysis:
    - Who uses the interface? What are their stress levels, time constraints, and cognitive loads?

  - What decisions are they making based on the interface?
- Identify critical data touchpoints and failure points in historical operations.

Step 2: Design for System Transparency

- Ensure interfaces clearly communicate:
  - Current system state (e.g., operational, standby, degraded, and error).
  - Status of key components (e.g., "sensor active," "calibration expired," and "backup engaged").
  - Whether a value is measured, estimated, or stale.
- Use color coding, status icons, and tooltips to make system states unambiguous.

Example: Instead of "Temp = 85°F," show "Temp = 85°F (Last updated 7 min ago, sensor in 'Warning' state)".

Step 3: Highlight Anomalies and Trends, Not Just Static Readouts

- Include:
  - Visual alerts for out-of-range values or trend deviations.
  - Time-series context to show whether a value is stable, rising, or erratic.
  - Confidence indicators or uncertainty shading for model outputs.
- Avoid overwhelming users with unnecessary noise design alert tiers by severity and actionability.

Step 4: Design for Clarity Under Pressure

- Use:
  - Simple, high-contrast layouts
  - Consistent iconography and data groupings
  - Minimized text during high-alert scenarios
- Provide "at-a-glance" summaries alongside deep drill-downs so that:
  - Operators can act quickly
  - Analysts can investigate further
  - Executives can assess overall status

Step 5: Build Interfaces That Support Skepticism and Diagnosis

- Include features that encourage users to question or investigate data, such as:
  - "Why is this value here?" buttons with supporting data lineage or change history.
  - Rollovers or expandable panels that show assumptions, input sources, and model logic.
  - Side-by-side comparison tools or "flag for review" options when something seems off.

Step 6: Train Users on System Operation and System Fallibility

- Move beyond "how to operate" toward:
  - How to detect drift, failure, or unexpected behavior.
  - Recognizing signs of stale, miscalibrated, or misclassified data.
  - Validating results using secondary cues or external verification tools.
- Include scenario-based simulations in training (e.g., "The sensor is lying what would you do?").

Step 7: Test Interfaces Under Realistic Conditions

- Conduct usability testing under time pressure, fatigue, and simulated alerts.
- Use A/B testing and cognitive walkthroughs to identify confusing or misleading elements.
- Incorporate feedback from diverse user roles (e.g., technicians, engineers, and managers).

Cultural and Organizational Enablers

- Promote a mindset that encourages data questioning, not just compliance.
- Reward users who catch interface errors or suggest improvements.
- Treat interface usability as a critical component of risk management, not just design.

Summary of Benefits

Practice	Benefit
Clear system state indicators	Reduces ambiguity and decision latency
Visual anomaly detection	Helps catch silent or slow-developing system drift
Transparent data context	Builds trust and enables better diagnosis
Scenario-based training	Prepares users for real-world uncertainty and failure
Human-centered layout design	Supports better decision-making under stress

A Real-World Example

In the Three Mile Island nuclear incident, a light on the control panel indicated that a relief valve had been "closed" but it only showed that the signal to close had been sent, not whether the valve had physically closed. The interface failed to convey the true system state, contributing to a misdiagnosis and partial core meltdown.

Had the interface included true/false valve position confirmation, or highlighted the discrepancy between expected and actual system behavior, the outcome may have been dramatically different.

Prioritizing human factors in interface design is about respecting the cognitive limits of users and the complexity of the systems they manage. It is not just about making things easier to use it is about making them harder to misuse, especially when the stakes are highest. A well-designed interface does not just display data it communicates truth, invites scrutiny, and guides wise action.

15

More Steps for Institutionalizing Retrospective Analysis

Abstract
Conduct blameless postmortems after major projects or close calls to identify silent risks and establish a "Lessons Learned Register" to integrate findings into training programs. This approach transforms failures into learning opportunities, enhances systemic insight, and prevents recurrence. Normalize skepticism and inquiry in operations, rewarding whistleblowers and dissenters. Use simulations and scenario testing to validate system resilience beyond historical data. Prioritize human factors in interface design, ensuring clarity and anomaly detection. Promote cross-functional decision-making, and formalize change detection systems. These practices build a culture of continuous improvement, transparency, and adaptive learning in high-reliability organizations.

Keywords: Patient safety, diagnostic accuracy, medical devices, treatment efficacy, clinical standards, health outcomes, compliance

After major projects or close calls, conduct blameless postmortems to surface silent risks. Establish a "Lessons Learned Register" tied to training programs.

In high-reliability organizations, failure is not just an outcome; it is a learning opportunity. Yet in many environments, failures and near misses are either swept under the rug or analyzed through a lens of blame, limiting the potential for meaningful improvement. To move beyond this, organizations must institutionalize retrospective analysis, a systematic, non-punitive process that examines events not only to assign responsibility but also to identify silent risks, process weaknesses, and cultural blind spots.

When done well, retrospective analysis becomes more than a reactive tool; it becomes a core component of continuous improvement, feeding lessons directly into training, design decisions, risk assessments, and even strategy.

Michael D. Holloway. Accuracy and Calibration Failures in Science and Engineering, (199–218)

Step-by-Step Process to Institutionalize Retrospective Analysis

Step 1: Define Triggers for Postmortems

- Formalize a policy that mandates retrospective reviews after:
  - Major project completions
  - Safety or quality incidents
  - "Near misses" (events that could have caused harm but did not)
  - Unexpected successes (to understand what worked)
- Include clear thresholds (e.g., downtime > 12 hrs, financial loss > $100K, and public impact) as automatic triggers.

Step 2: Establish a Blameless Postmortem Framework

- Use a structured template to guide postmortems, including:
  - Timeline of events
  - Key decisions and data inputs at each stage
  - Contributing factors (technical, procedural, human)
  - Latent conditions (e.g., unclear SOPs, outdated training, and ambiguous interfaces)
  - Systemic improvements identified
- Emphasize psychological safety:
  - State clearly that the purpose is learning, not punishment.
  - Facilitate with a neutral moderator.
  - Focus on what happened and why, not who to blame.

Step 3: Surface Silent and Systemic Risks

- Go beyond the immediate root cause:
  - Ask what allowed the issue to happen without being caught.
  - Examine underlying assumptions, poor feedback loops, or outdated legacy knowledge.
  - Consider cultural factors, e.g., hierarchy suppression, deadline pressure, and normalization of deviance.

Example: If a sensor failed, ask why no one noticed. Was the alarm buried? Was there no training on drift detection?

Step 4: Create a Centralized "Lessons Learned Register"

- Log postmortem findings in a searchable, well-organized Lessons Learned Register, including:
  - Date, event type, summary
  - Recommendations implemented (and their status)
  - Links to SOPs or system changes
  - Tags for cross-referencing (e.g., "interface design," "training gap," and "miscalibrated model")
- Use this register to:
  - Prevent recurrence across teams or projects
  - Feed future risk assessments
  - Enhance project planning and scenario simulations

Step 5: Tie Lessons Directly to Training and SOPs

- Review postmortem findings regularly to:
  - Update training curricula
  - Revise SOPs and design checklists
  - Create scenario-based drills using real past incidents
- Integrate these findings into:
  - New employee onboarding
  - Recertification or annual training
  - Cross-functional workshops

Step 6: Evaluate and Improve the Process

- Survey participants after each postmortem:
  - Was the process constructive and respectful?
  - Did it generate actionable insights?
  - Were follow-ups completed and communicated?
- Continuously refine the structure and facilitation of retrospectives based on feedback.

Cultural and Leadership Enablers

- Leadership must model humility and be willing to review mistakes and openly share what they learned.

- Celebrate the uncovering of "invisible failures", not just successful outcomes.
- Recognize individuals who flag near misses or challenge accepted practices; they are often your early warning system.

Practice	Benefit
Blameless postmortems	Promotes honesty, transparency, and systemic insight
Central Lessons Learned Register	Prevents institutional memory loss and siloed learning
Training integration	Makes insights actionable and prevents recurrence
Feedback loop to SOPs	Ensures continuous operational refinement

Example in Practice
After the Mars Climate Orbiter loss, a retrospective would have uncovered:

- Lack of unit consistency checks
- Insufficient cross-team interface validation
- A culture of assumption instead of verification

These insights could have been translated into:

- Mandatory unit declarations in all data files
- Cross-functional checklist requirements for interface control documents
- Updated training for mission software engineers

Institutionalizing retrospective analysis transforms past failures into future resilience. It ensures that the cost of mistakes is not only wasted but also reinvested into smarter systems, better training, and a stronger culture. In a learning organization, the question is not "Who's at fault?" but "What can we learn, and how can we share it?" Because when hindsight becomes foresight, every close call becomes a catalyst for improvement.

Maintain Healthy Skepticism and a Culture of Inquiry

- Promote a mindset that welcomes questioning and doubt, even during confident operations.
- Protect and reward whistleblowers, dissenters, and truth-seekers.

In complex, high-consequence systems, overconfidence is often the silent prelude to failure. When individuals or organizations assume that procedures are foolproof, that models are infallible, or that "the system knows best," they create the very conditions in which blind spots grow. To combat this, organizations must actively cultivate a culture of inquiry, one that values questioning, critical thinking, and the courage to speak up, especially when operations appear to be running smoothly.

Healthy skepticism is not about cynicism or paralysis; it is about ensuring that confidence is continually tested against evidence, logic, and evolving context. It is a self-correcting mechanism that, when embedded culturally, enables early detection of drift, uncovering of misinterpretations, and adaptation to change.

Step-by-Step Process to Maintain Healthy Skepticism and a Culture of Inquiry
Step 1: Explicitly Define and Normalize Skepticism

- Include "healthy skepticism" as a core value in company mission statements or quality principles.
- Make it clear that questioning the system is part of good operations, not an act of disloyalty or inefficiency.
- Communicate that no model, process, or dashboard is immune to scrutiny, especially the most trusted ones.

Step 2: Train for Critical Thinking and Cognitive Bias Recognition

- Offer structured training programs that teach:
    - How to spot confirmation bias, groupthink, and assumption traps.
    - Techniques for Socratic questioning, devil's advocacy, and root cause interrogation.
    - How to interpret data with uncertainty, rather than defaulting to single-point conclusions.

- Use real-world case studies where skepticism would have prevented failure.

Step 3: Build Inquiry Into Operational Routines

- Integrate "challenge questions" into SOPs, project reviews, and risk assessments:
  - "What is the worst-case scenario we're not considering?"
  - "What assumption, if wrong, would undermine this conclusion?"
  - "What would someone from another department say?"
- Make inquiry part of normal conversations, not just emergency drills or retrospectives.

Step 4: Protect and Empower Dissenters

- Enact formal non-retaliation policies for employees who raise concerns, offer unpopular perspectives, or challenge accepted narratives.
- Establish confidential reporting channels and a designated advocate or ombudsman for those who fear reprisal.
- Regularly publicize and celebrate cases where internal dissent led to improvements, showing that dissent is valued, not punished.

Step 5: Reward Informed Dissent and Early Warnings

- Create mechanisms to recognize employees who identify weak signals, near misses, or unnoticed inconsistencies:
  - Internal awards, recognition boards, and bonuses tied to risk mitigation.
  - "Golden Flag" or "Truth Seeker" awards for those who stopped a bad decision before it reached execution.
- Acknowledge "the mistake we did not make" as an achievement.

Step 6: Foster Cross-Hierarchy Openness

- Train leaders and managers to invite skepticism openly:
  - "What might we be missing?"
  - "Is there a piece of bad news no one wants to say?"
  - "If this fails next month, why will it fail?"

- Flatten hierarchical structures during reviews and encourage junior staff and frontline workers to challenge assumptions.

Step 7: Monitor for Complacency and Overconfidence

- Establish cultural metrics in employee surveys:
  - "Do you feel safe raising concerns?"
  - "Are alternative viewpoints encouraged?"
  - "Does leadership respond constructively to disagreement?"
- Include skepticism health checks in organizational audits and leadership evaluations.

Cultural and Organizational Enablers

- Tone from the top: Executives must model humility and curiosity.
- Make space for "constructive friction" in teams.
- Promote a culture where being wrong is acceptable but failing to question is not.

Practice	**Benefit**
Normalized questioning	Prevents groupthink and model overreliance
Whistleblower protection	Enables early risk identification
Recognition of dissent	Builds psychological safety and trust
Embedded challenge routines	Drives continuous improvement and adaptive learning
Cross-level dialogue	Improves decision robustness through diverse perspectives

Example in Practice
During the Challenger launch decision, had dissenting engineers' concerns about cold-weather O-ring performance been welcomed and elevated rather than overruled under pressure, the disaster may have been averted. A culture of inquiry would have slowed the decision process, favoring truth over schedule.

Fostering healthy skepticism and a culture of inquiry is about building intellectual resilience into your organization's DNA. It ensures that trust in systems is earned, continually tested, and never blind. In this kind of environment, people are not just allowed to speak up; they are expected to. And when they do, they are not seen as troublemakers; they are seen as protectors of truth, integrity, and safety.

Use Simulation and Scenario Testing Regularly

- Run worst-case drills, simulations, and model perturbations to test system resilience.
- Validate models not just by fit to past data, but by performance under edge conditions.

Assumptions can quickly become liabilities if left untested. Organizations that rely solely on historical data or best-case operations are at risk of being blindsided by rare but high-impact events. That is why it is critical to regularly run simulations and scenario tests that explore how systems behave under stress, failure, uncertainty, and surprise.

Simulation and scenario testing enable organizations to go beyond theoretical reliability and actually experience how a system, model, or team performs when the environment becomes unpredictable. These exercises are not just about disaster preparedness; they are about resilience engineering, learning through controlled failure, and ensuring that models and plans hold up at the edges, not just in the middle.

Step-by-Step Process to Implement Regular Simulation and Scenario Testing

Step 1: Identify What Needs to Be Tested

- Prioritize high-impact systems and models:
  - Predictive algorithms driving safety or financial decisions
  - Critical infrastructure (e.g., power, process control, logistics, and data pipelines)
  - Emergency response protocols and crisis communication plans
- For each target, ask, "What could go catastrophically wrong, and how ready are we?"

Step 2: Define Realistic and Extreme Scenarios

- Develop a range of scenarios, including:
  - Baseline variations: sensor lag, minor input drift, normal system degradation
  - Edge conditions: max load, high latency, untested environments (temperature, altitude, etc.)
  - Worst-case and black swan events: complete system failure, cyberattack, model drift + data loss
  - Human-error events: misconfigured settings, incorrect assumptions, delayed response

Example: For a predictive maintenance model, simulate what happens if sensors are calibrated incorrectly or if the failure signal is masked by normal operating noise.

Step 3: Design the Simulation Methodology

- Choose appropriate simulation techniques:
  - Monte Carlo simulations for probabilistic risk
  - Digital twins for real-time system emulation
  - Tabletop exercises for crisis response scenarios
  - Perturbation testing for AI/ML models (e.g., adversarial inputs and stress inputs)
- Use synthetic datasets and data fuzzing to test model robustness.

Step 4: Conduct Scenario Walkthroughs with Cross-Functional Teams

- Include:
  - Technical personnel to validate system behavior
  - Operators to assess human decision paths
  - Executives to evaluate organizational readiness and communication flows
- Use pre-briefs and postmortems:
  - Pre-brief: Define objectives, rules, and metrics.
  - Postmortem: Document findings, decisions made, and capability gaps discovered.

Step 5: Evaluate Model Robustness Beyond Historical Fit

- Do not just assess if the model "fit the past"; evaluate how it performs when:
  - Inputs fall outside training bounds
  - Unexpected interactions occur between variables
  - Environmental context shifts (e.g., seasonality, pandemic, and policy change)
- Compare results across multiple models or versions, and measure stability, error tolerance, and confidence drop-off.

Step 6: Capture Lessons and Feed Back Into Systems

- After each simulation or drill:
  - Update SOPs, dashboards, and training materials.
  - Tag gaps in data collection, system responsiveness, or decision quality.
  - Prioritize design or policy changes based on simulation outcomes.
- Feed outcomes into:
  - The Lessons Learned Register
  - Risk heat maps
  - Future simulation design (build increasingly challenging rounds)

Step 7: Institutionalize and Schedule Simulation Cadence

- Build simulation testing into operational calendars:
  - Monthly microdrills
  - Quarterly model stress tests
  - Annual full-system failure simulations
- Make simulation results required review materials for audits, certifications, and strategic planning.

Cultural and Organizational Enablers

- Shift the narrative: Simulated failure is a sign of maturity, not weakness.

- Empower staff to treat drills seriously and encourage creativity and realism in scenario design.
- Avoid post-simulation blame. Focus on systems thinking and process gaps, not personal fault.

Practice	**Benefit**
Scenario diversity	Exposes blind spots and stress vulnerabilities
Edge-condition testing	Ensures the robustness of models and systems in unfamiliar contexts
Crisis drills	Prepares staff for real-world response and communication
Feedback loops	Drives continuous system and model improvement
Institutional cadence	Makes preparedness a habit, not a reaction

Example in Practice

In the aviation industry, simulators regularly expose pilots to emergency scenarios they've never encountered in flight, such as engine failure after takeoff or sudden weather disruptions. These drills do not just teach procedures; they embed instincts, ensuring pilots respond correctly even under stress.

Similarly, cybersecurity teams run "red team" penetration tests to simulate system breaches and measure response. Without such drills, systems that appear secure may be dangerously vulnerable.

Simulation and scenario testing transform theory into experience. They reveal where resilience is real and where it is only assumed. By testing systems and people under stress, organizations build the muscle memory, situational awareness, and design maturity needed to thrive in complexity and survive the unexpected. In today's world, it is not enough for a system to work well; it must fail gracefully, recover quickly, and improve continuously.

"Complex systems fail when we mistake the illusion of certainty for real control. The antidote is a culture of transparency, challenge, and resilience."

Categorized Root Causes and Corrective Actions for Historical Incidents

Incident	Root cause category	Corrective actions/safeguards
Challenger Disaster	Data misinterpretation & communication failure	Mandate data visualization clarity and direct engineering-to-executive communication during critical decisions.
Chernobyl Disaster	Procedural deviation & inadequate training	Enforce protocol compliance audits and train operators on system behavior under edge conditions.
Three Mile Island	Misinformation from Interface Design	Redesign the interface to better reflect the true system state; add alarm prioritization and automated diagnostics.
COVID-19 Initial Response	Incomplete data & delay in response	Improve global health data transparency, build real-time epidemiological dashboards, and run simulations regularly.
Mars Climate Orbiter	Data standardization failure	Standardize data formats and units across international teams; validate all mission-critical software inputs.
2008 Financial Crisis	Faulty risk models & data omission	Stress test financial models with real-world variability; demand transparency in risk assumptions and data sources.
Fukushima Daiichi Disaster	Underestimated Risk from Historical Data	Incorporate extreme scenario planning; update hazard models using full historical and geological data.
Y2K Panic	Overcompensated perceived risk	Use tiered threat evaluation; balance readiness with data-driven risk communication to avoid overreaction.

Philosophical and Ethical Dimensions

Philosophical and Ethical Dimensions of Denial and Error Suppression in Technical Systems

At the heart of many calibration-related failures lies not just a technical oversight, but a deeply human and institutional reluctance to confront uncomfortable truths. The psychology of denial plays a powerful role in the persistence of flawed data. Cognitive biases such as normalcy bias, confirmation bias, and motivated reasoning can lead engineers, scientists, and administrators to discount anomalous data that challenges institutional narratives or operational expectations (Kahneman, 2011). This denial is not necessarily malicious; rather, it often stems from a desire to maintain stability, avoid disruption, and preserve professional identity. When the data itself becomes suspect, the human tendency is often to question the instrument before questioning the system it undermines.

This psychological tendency is compounded by institutional resistance to admitting error, particularly in high-stakes industries where liability, public trust, and market position are at risk. Large organizations often develop internal cultures that penalize whistleblowing and reward conformity, reinforcing systems of silence even when data integrity is compromised (Vaughan, 1996). Bureaucratic inertia and hierarchical decision-making can render it easier to discredit a failed calibration or misaligned measurement than to reassess foundational assumptions. In some cases, such as Volkswagen's Dieselgate scandal, the response was active deception: engineers deliberately manipulated calibration protocols to hide NOx emissions from regulators. In others, like the NASA Challenger disaster, the warning signs were known but suppressed through rationalized groupthink and management bias (Rogers Commission, 1986).

Conversely, some industries and institutions have demonstrated proactive and ethical responsiveness when faced with drift or calibration uncertainty. In semiconductor manufacturing, companies like Intel and TSMC have embraced constant recalibration and metrology at the atomic scale, embedding error detection and self-correction as part of standard process control (Ho & O'Neill, 2020). The metrology community's response to the drift of the International Prototype Kilogram stands as a model of intellectual humility and scientific responsibility, opting to redefine the unit through Planck's constant after acknowledging the limitations of artifact-based measurement (BIPM, 2019). In clinical diagnostics, laboratories accredited under ISO 15189 and CLSI guidelines have begun instituting automated alerts and statistical controls to detect calibration drift in IVD devices, exemplifying a data-informed, ethically grounded culture of safety and traceability (CLSI, 2018).

These contrasting responses illuminate a key philosophical divide: whether an institution treats error as an opportunity for epistemic growth and adaptive learning or as a threat to be concealed. The ethical imperative, particularly in safety-critical domains, is to design systems not just for robustness but for transparency and humility, recognizing that even the most advanced instruments are only as trustworthy as the organizational cultures that monitor, interpret, and act upon their outputs. A commitment to calibration integrity, therefore, is not merely technical; it is moral.

The Psychology of Denial in the Face of Flawed Data

Denial in the presence of flawed or contradictory data is not simply an act of ignorance or willful blindness; it is often a psychologically motivated defense mechanism deeply embedded in human cognition and organizational culture. At its core, denial allows individuals and institutions to protect their existing mental models, values, and sense of competency from dissonant information that may challenge them. This tendency is frequently supported by cognitive biases such as confirmation bias, where individuals unconsciously favor data that supports their expectations, and normalcy bias, where they underestimate the possibility of deviation or failure in familiar systems (Kahneman, 2011; Tversky & Kahneman, 1974). In complex systems, especially in science, engineering, and high-risk operations, this psychological armor can become dangerous, as it leads to the dismissal of early warning signs, the rationalization of anomalies, and the selective interpretation of results to maintain operational or institutional continuity.

A classic example of this dynamic can be seen in Diane Vaughan's seminal study of the Challenger space shuttle disaster, where flawed O-ring data was known and repeatedly downplayed. Vaughan termed this phenomenon "the normalization of deviance", in which engineers and managers reinterpreted failed data points as an acceptable risk over time (Vaughan, 1996). In this way, flawed data does not immediately provoke crisis; instead, it is absorbed and recontextualized to avoid disrupting workflow, timelines, or reputational capital. This behavior reflects a broader human discomfort with ambiguity and error, a discomfort that is amplified in hierarchical or high-performance environments where admitting uncertainty may be seen as weakness or incompetence (Janis, 1982).

The psychology of denial is also reinforced by group dynamics, especially when individuals operate in insular or tightly knit teams. Groupthink, a term coined by Irving Janis, describes how the desire for consensus can suppress dissenting voices, discourage scrutiny of data, and create illusions of infallibility. In such environments, even individuals who detect flaws in data may self-censor or rationalize discrepancies, believing they must be mistaken if others are unconcerned. This suppression is not always overt; it can take the form of motivated forgetting or attentional filtering, where inconvenient information is simply not retained or processed with the same seriousness as confirmatory data (Festinger, 1957; Tavris & Aronson, 2007).

Modern research in behavioral science also highlights the concept of "epistemic closure", where individuals operate within sealed belief systems that selectively admit or reject data based on ideological, cultural, or professional boundaries (Sunstein, 2009). In technical fields, this can manifest as overconfidence in instruments, models, or reputational authority, where institutional prestige substitutes for empirical rigor. When calibration data or measurement trends contradict expectations, denial becomes a convenient tool to preserve institutional narratives and avoid resource-intensive investigations.

Ironically, the more specialized and data-driven an organization becomes, the greater the potential for denial to take root because high complexity provides ample cover for interpretive flexibility and *post-hoc* rationalization. As noted by sociologist Charles Perrow in *Normal Accidents*, tightly coupled systems often leave little room for trial and error, making denial a preferred path to avoid triggering systemic disruption (Perrow, 1984). Thus, the psychology of denial in the face of flawed data is not simply a failure of logic; it is a defense of identity, order, and belief—a defense that must be consciously dismantled through cultural reform, transparency, and the systematic design of error-tolerant systems.

References

- Festinger, L., *A Theory of Cognitive Dissonance*, Stanford University Press, 1957.
- Janis, I., *Groupthink: Psychological Studies of Policy Decisions and Fiascoes*, Houghton Mifflin, 1982.
- Kahneman, D., *Thinking, Fast and Slow*, Farrar, Straus and Giroux, 2011.
- Perrow, C., *Normal Accidents: Living with High-Risk Technologies*, Princeton University Press, 1984.

- Sunstein, C.R., *Going to Extremes: How Like Minds Unite and Divide*, Oxford University Press, 2009.
- Tavris, C. and Aronson, E., *Mistakes Were Made (But Not by Me)*, Harcourt, 2007.
- Tversky, A. and Kahneman, D., Judgment under Uncertainty: Heuristics and Biases. *Science*, 185, 4157, 1124–1131, 1974.
- Vaughan, D., *The Challenger Launch Decision: Risky Technology, Culture, and Deviance at NASA*, University of Chicago Press, 1996.

Institutional Resistance to Admitting Error

Institutional resistance to admitting error is not merely a bureaucratic delay; it is often a deeply embedded cultural, structural, and psychological phenomenon. In many organizations, particularly those with high public visibility, technical authority, or economic power, the acknowledgment of error carries significant reputational, legal, and financial consequences. This creates an environment where maintaining the illusion of infallibility becomes more important than investigating or correcting mistakes. As Argyris (1991) notes in his work on organizational learning, many institutions operate within a "Model I" culture, where defensive routines, blame avoidance, and face-saving behaviors dominate, making double-loop learning, where underlying assumptions are questioned, rare or unwelcome. In such systems, errors are often buried, reframed as outliers, or attributed to externalities rather than acknowledged as systemic failures.

The consequences of this resistance are well-documented. The Columbia space shuttle disaster in 2003, like its Challenger predecessor, revealed a culture of silencing dissent and normalizing abnormal data. Engineers who raised concerns about foam strike damage were dismissed or overruled, and safety review processes were undermined by hierarchical pressure and risk minimization (CAIB, 2003). Similarly, in the Volkswagen emissions scandal, the company went to great lengths to obscure deliberate calibration manipulation rather than confront the ethical and regulatory implications of its defeat device software, ultimately resulting in criminal penalties and massive brand damage. What both cases illustrate is that hierarchical organizations often suppress truth-telling when it threatens authority and build internal mechanisms that discourage the escalation of inconvenient findings (Vaughan, 1996; Schein, 2010).

This resistance is further amplified by what sociologists call "institutional isomorphism," a tendency of organizations to mimic one another in structure and practice to maintain legitimacy (DiMaggio & Powell, 1983). Once denial or downplaying of error becomes normalized within a sector (e.g., aviation, finance, and medicine), it becomes institutionally reinforced. Leaders may fear being outliers by openly admitting fault, especially if competitors remain silent. Furthermore, regulatory capture, where oversight bodies develop too close a relationship with the industries they regulate, can lead to mutual interest in obscuring errors that would otherwise disrupt economic or political equilibrium (Stigler, 1971).

Overcoming institutional resistance requires more than policy change; it demands a cultural commitment to transparency, psychological safety, and system-level learning. Organizations such as the U.S. National Transportation Safety Board (NTSB) and aviation maintenance operations under SMS frameworks have made strides by embedding independent review, root cause analysis, and non-punitive reporting into their systems. Yet even in such environments, resistance can persist if leadership fails to model accountability or if whistleblowers are not protected. Ultimately, institutional resistance to admitting error is less about ignorance and more about self-preservation, a dynamic that must be addressed not only with technical checks but with ethical leadership and structural redesign that rewards openness rather than punishes it.

References

- Argyris, C., Teaching Smart People How to Learn. *Harv. Bus. Rev.*, 69, 3, 99–109, 1991.
- Columbia Accident Investigation Board (CAIB), *Report Volume I: The Columbia Accident Investigation Board Report*, NASA, 2003.
- DiMaggio, P.J. and Powell, W.W., The Iron Cage Revisited: Institutional Isomorphism and Collective Rationality in Organizational Fields. *Am. Sociol. Rev.*, 48, 2, 147–160, 1983.
- Schein, E.H., *Organizational Culture and Leadership*, Jossey-Bass, 2010.
- Stigler, G.J., The Theory of Economic Regulation. *Bell J. Econ. Manage. Sci.*, 2, 1, 3–21, 1971.
- Vaughan, D., *The Challenger Launch Decision: Risky Technology, Culture, and Deviance at NASA*, University of Chicago Press, 1996.

Responsiveness vs. Obfuscation in the Face of Flawed Data

While flawed data and calibration drift are widespread challenges, industry responses vary dramatically, often defining the boundary between systemic resilience and catastrophic failure. In aerospace and semiconductor manufacturing, for example, sectors characterized by tight tolerances and catastrophic risk profiles, many organizations have responded proactively to calibration uncertainties. Companies like Intel, ASML, and TSMC have implemented sub-nanometer metrology controls, redundant process checks, and closed-loop feedback systems to detect and correct drift before it impacts yield or safety (Ho & O'Neill, 2020). Similarly, the U.S. National Transportation Safety Board (NTSB) and segments of the aviation industry operating under Safety Management Systems (SMS) have embraced just culture policies, black box data transparency, and post-incident root cause analysis as a baseline for trust and improvement. These models exhibit a shared ethic: that error exposure is essential to learning and that resilience is built on the back of transparent calibration and verification routines.

In contrast, industries with complex regulatory entanglements or strong market incentives to preserve their image have historically exhibited institutional resistance and cover-up behavior when facing similar flaws. The automotive industry, in the case of Volkswagen's Dieselgate, represents a deliberate burial of calibration manipulation where engineers coded emissions control software to cheat regulatory tests while falsely claiming environmental compliance (U.S. DOJ, 2017). Similarly, Theranos built its brand on the fiction of technological precision, simulating calibration confidence while obfuscating its devices' inability to produce reproducible results—an institutional fraud justified internally through secrecy and selective validation (Carreyrou, 2018). Even in healthcare, the early months of the COVID-19 pandemic saw mismanagement of PCR testing protocols and cycle threshold calibration inconsistencies, where public health institutions hesitated to standardize or acknowledge procedural flaws in test interpretation, thus distorting case rates and public risk perceptions (Ioannidis, 2020).

The key distinction between responsive and resistant sectors is often cultural: responsive industries foster distributed ownership of quality, invest in metrological literacy, and institutionalize feedback mechanisms that elevate anomalies rather than suppress them. Resistant industries, on the other hand, tend to centralize authority, punish whistleblowing, and treat flawed data as reputational threats rather than operational warnings. These comparisons reveal that calibration integrity is not only technical; it is deeply cultural, ethical, and strategic. The willingness to confront uncomfortable data is now a defining metric of long-term organizational health.

References

- Carreyrou, J., *Bad Blood: Secrets and Lies in a Silicon Valley Startup*, Knopf, 2018.

- Ho, M. and O'Neill, B., Metrology for the Next Generation: Atomic-Level Calibration in Semiconductor Fabrication. *J. Microelectron. Manuf.*, 45, 2, 58–66, 2020.
- Ioannidis, J.P.A., Precision and Bias in COVID-19 Test Interpretation. *Eur. J. Clin. Invest.*, 50, 6, e13222, 2020.
- U.S. Department of Justice, *Volkswagen AG Agrees to Plead Guilty and Pay $4.3 Billion in Criminal and Civil Penalties*, 2017.
- Vaughan, D., *The Challenger Launch Decision: Risky Technology, Culture, and Deviance at NASA*, University of Chicago Press, 1996.

When the Truth is Inconvenient – Human and Institutional Responses to Flawed Data

This chapter has examined the deeper human and organizational dynamics that surround failures in calibration, measurement, and data integrity not only as technical oversights but also as expressions of cognitive bias, cultural inertia, and ethical fragility.

We began by exploring the psychology of denial, where flawed or anomalous data is often filtered through subconscious mechanisms of self-protection. From confirmation bias to the normalization of deviance, individuals unconsciously reinterpret or disregard warning signs to preserve their sense of control, competence, or continuity. In technical systems, this psychological vulnerability can become especially dangerous when it interacts with high complexity and ambiguous accountability.

At the organizational level, we explored institutional resistance to admitting error, a phenomenon driven by reputational risk, liability exposure, and fear of disruption. Large institutions often develop systems of silence, where defensive routines, hierarchical suppression, and performance pressure lead to the burying of inconvenient truths. Such environments may rationalize drift, delay corrective action, or even manipulate the narrative, choosing image over insight.

Finally, we compared how different industries have responded to these pressures. In aerospace, semiconductor fabrication, and regulated laboratories, we found constructive cultures that embraced transparency, embedded feedback systems, and institutionalized recalibration. In contrast, cases like Volkswagen's emissions fraud, Theranos' pseudocalibration, and early PCR test mismanagement during the COVID-19 crisis reveal the ethical collapse that occurs when organizations choose concealment overcorrection.

Across all cases, one theme stands firm: calibration is not just a technical responsibility; it is a philosophical and ethical stance. To preserve the integrity of data, systems must be designed not just for accuracy, but for accountability, humility, and learning. Only by recognizing that denial and resistance are natural but not inevitable can institutions move from fragility to resilience and from error to insight.

References

- BIPM (Bureau International des Poids et Mesures). *The International System of Units (SI) (9th ed.)*. BIPM, 2019. ISBN: 978-9292212491.
- Clinical and Laboratory Standards Institute (CLSI). *C29-A3: Calibration and quality control of sodium and potassium ion-selective electrode measurements*, CLSI, 2018.
- Ho, M. and O'Neill, B., Metrology for the next generation: Atomic-level calibration in semiconductor fabrication. *J. Microelectron. Manuf.*, 45, 2, 58–66, 2020, https://doi.org/10.1016/j.jmmb.2020.01.002.
- Kahneman, D., *Thinking, fast and slow*. Farrar, Straus and Giroux, 2011. ISBN: 978-0374275631.
- Rogers Commission. *Report of the Presidential Commission on the Space Shuttle Challenger Accident*. U.S. Government Printing Office, 1986.
- Vaughan, D., *The Challenger Launch Decision: Risky Technology, Culture, and Deviance at NASA*, University of Chicago Press, 1996.
- U.S. Department of Justice, Volkswagen AG Agrees to Plead Guilty and Pay $4.3 Billion in Criminal and Civil Penalties, 2017, https://www.justice.gov.

16

Implementation & Culture Change – Embedding Calibration Integrity Across the Enterprise

Abstract

Technical accuracy alone cannot ensure data integrity without a culture that values questioning and accountability. Building an organization-wide calibration integrity culture requires embedding transparency, traceability, and trust into all levels of operation. Leadership must model calibration as a strategic priority, fostering psychological safety for open communication. Shared vocabulary and metrics, training in metrological literacy, and integration with quality and risk programs reinforce this culture. Calibration management systems enhance visibility and accountability. Learning from failures and maintaining vigilance against data drift are essential. This transformation turns calibration from a compliance task into a strategic pillar of resilience and excellence.

***Keywords*:** Flight safety, navigation systems, avionics, structural integrity, performance testing, certification, reliability

The most sobering insight from the failures documented in previous chapters is that technical accuracy alone is not enough. No algorithm, instrument, or standard, however precise, can prevent drift, distortion, or failure when embedded within a culture that tolerates silence, punishes questioning, or disconnects data from accountability. The next frontier in data integrity is not technological; it is CULTURAL.

This section confronts the central challenge of operational transformation: how to build a system-wide culture of calibration integrity, one that recognizes measurement not as a discrete event, but as an ongoing, organization-wide commitment to transparency, traceability, and trust. From quality assurance laboratories to field operations, from executive leadership to front-line technicians, the success of any calibration program depends not only on what is measured but also on how measurement is valued, verified, and acted upon.

Michael D. Holloway. Accuracy and Calibration Failures in Science and Engineering, (219–228)

We begin by capturing the voices of those most responsible for data reliability: QA/QC managers, calibration engineers, and lab directors, whose real-world experiences reveal both the vulnerabilities and breakthroughs of calibration systems in practice. Their insights reflect the challenges of gaining buy-in, implementing system-wide procedures, and moving from a compliance mindset to a resilience mindset.

From there, we present examples of standard operating procedures (SOPs), digital dashboard frameworks, and audit checklists that reflect best practices in calibration governance. These tools demonstrate what "good" looks like not as rigid compliance artifacts, but as living documents that foster early warning, learning loops, and measurement transparency.

To guide organizations in assessing and advancing their readiness, we introduce a Calibration Culture Maturity Model, also called the Measurement Integrity Readiness Index (MIRI). This model benchmarks institutions across a continuum from reactive and fragmented to predictive and integrated, offering a structured path to improvement. Organizations at higher levels of maturity do not merely prevent failure; they cultivate the ability to detect, question, and adapt before harm occurs.

Ultimately, this chapter is about transformation: not just fixing instruments, but reforming institutions. It is about turning what was once a hidden technical afterthought calibration into a strategic pillar of operational excellence, ethical responsibility, and system resilience.

Building an Organization-Wide Calibration Integrity Culture

Creating a resilient organization that can withstand data drift, instrument failure, and measurement uncertainty requires more than periodic calibration; it demands a **culture of calibration integrity** that is embedded into every level of operations. Calibration must cease being seen as an isolated technical task performed by specialists and instead become a **shared institutional value**, understood and supported by technicians, managers, executives, and regulators alike.

From Compliance to Commitment

In many organizations, calibration exists within a **compliance framework,** a checkbox in an audit trail, or a scheduled maintenance activity. While meeting ISO/IEC 17025 or FDA 21 CFR Part 11 requirements is necessary, **compliance alone is not culture**. A calibration integrity culture is one in which employees recognize that **every decision made using flawed data, whether in production, diagnostics, or design, can undermine product quality, safety, and public trust**. This recognition must be **proactively cultivated**, not retroactively enforced.

To do so, calibration must be reframed as **a critical control point in organizational decision-making**. Every department, from procurement and R&D to maintenance and quality control, must understand its reliance on verified, traceable measurements. Organizations must consistently ask: *What assumptions are embedded in our data? What is the health of the instruments that generate them?* When these questions become routine, culture change begins.

Leadership and Psychological Safety

Culture begins at the top. Executive leaders and senior technical managers must **model calibration integrity as a strategic priority**, not just a technical concern. This includes **allocating resources**, **staffing metrology teams adequately**, and **publicly responding to data integrity issues** without blame or suppression. Where data uncertainty arises, leaders should encourage inquiry, not evasion.

Equally important is **psychological safety**, the ability of employees to report, question, or escalate calibration concerns without fear of reprisal. Many past failures, from the Deepwater Horizon disaster to Theranos, were exacerbated not only by the absence of calibration systems but also by the **fear of speaking up** when those systems failed. A true calibration culture encourages vigilance, curiosity, and the surfacing of anomalies as a **source of strength, not weakness**.

Systemic Communication and Shared Vocabulary

Organizations often falter not because they lack good intentions, but because **they lack a common language**. Calibration, uncertainty, traceability, drift, and tolerance must be demystified across departments. Clear internal definitions and shared metrics such as Measurement System Analysis (MSA), uncertainty budgets, and error propagation should be embedded into workflows, dashboards, and cross-functional meetings.

This can be reinforced through **visual management tools**: real-time dashboards showing instrument status, color-coded calibration intervals, or trend warnings can make abstract concepts visible. SOPs should include not only technical instructions but also **explanations of why the calibration step matters**, what happens if it's skipped, and how it connects to product safety or regulatory compliance.

Training, Mentorship, and Lifelong Literacy

Sustained cultural change depends on the **development of metrological literacy at all levels**. Technicians, operators, engineers, and even executives need training not only in how to use calibrated tools but also in how to **critically assess data quality, interpret uncertainty, and understand drift behaviors**. This involves both **technical training** (e.g., understanding ISO/IEC 17025, ASTM standards, and calibration certificates) and **contextual training** (e.g., historical failure cases, and ethical obligations).

Organizations should implement **onboarding modules for new hires**, **annual calibration integrity refreshers**, and **mentorship systems** that connect junior staff with seasoned metrologists. Continuous professional development ensures that calibration integrity is not static; it evolves with technology, standards, and organizational learning.

Integration with Quality, Risk, and Ethics Programs

A calibration culture must be **embedded within the organization's broader quality, safety, and ethical frameworks**. For example:

- Quality Management Systems (QMS) should include **calibration-related KPIs** such as overdue calibrations, drift flags, and calibration nonconformities.
- Risk registers should assess the **impact of measurement uncertainty or calibration lapse** on critical operations.
- Ethical guidelines should explicitly state the obligation to **report and act on flawed data**, regardless of hierarchy or timeline pressure.

Calibration culture becomes institutionalized when it is **tied to performance reviews, audit scores, project approvals, and leadership evaluations**, not simply seen as a support function.

Data Systems, Automation, and Transparency

A modern calibration culture must be supported by **digital systems that enable visibility, traceability, and accountability**. Calibration management systems (CMS) should track:

- Instrument history and recalibration schedules
- Calibration status flags on live data
- User overrides and manual entries
- Audit logs for who accepted or rejected out-of-tolerance instruments

Where feasible, **automation should reduce dependence on memory or guesswork**. For example, integration of calibration status into LIMS (Laboratory Information Management Systems), ERP platforms, or SCADA environments can alert users to drift before it affects production or reporting.

Institutional Learning from Failure

Finally, organizations must turn **calibration errors into learning opportunities**. Every nonconformance or drift event should trigger not just a correction, but a **root cause investigation and systems-level review**. Post-mortems and cross-departmental calibration summits can build institutional memory and clarify where communication, training, or accountability broke down. Sharing lessons learned across departments or sites reinforces the idea that **measurement integrity is everyone's business**.

Building an organization-wide calibration integrity culture is not an initiative; it is an evolution. It requires **executive support, cross-functional collaboration, continuous training, and system-wide transparency**. It thrives in organizations where **data is sacred, curiosity is protected, and truth is more valuable than convenience**. When calibration is no longer viewed as a compliance burden but as a **pillar of ethical, scientific, and operational excellence**, the organization becomes more than compliant; it becomes resilient.

Interview Excerpts with QA/QC Managers and Lab Directors: Voices from the Front Lines of Calibration Integrity

To understand how calibration integrity is built or eroded within real organizations, it is essential to hear from those responsible for managing it daily. Interviews with **quality assurance (QA) managers**, **quality control (QC) supervisors**, and **laboratory directors** reveal the tension between compliance requirements, operational pressure, and cultural values. Their perspectives illuminate not only technical concerns but also **leadership, communication, and institutional accountability**.

Risk of Quiet Drift

> *"It's not the big failures that scare me; it's the slow drift that goes unnoticed until we're calibrating against our own assumptions."*
>
> *Senior Lab Director, Environmental Testing Facility, U.S.*

This quote underscores the insidious nature of calibration drift, particularly in labs that handle high-volume, repetitive testing. Without vigilant baseline checks and reference standards, labs may **calibrate new data against unverified trends**, leading to the normalization of error. Studies in analytical chemistry and process control confirm that **undetected drift can persist for months**, especially in systems lacking inter-laboratory verification (ASQ, 2020; ISO/IEC 17025:2017).

Cultural Resistance and Psychological Safety

> *"Nobody wants to be the one to stop production because a scale might be off. But if you don't empower people to raise the flag, you're gambling with every batch."*
>
> *QA Manager, Pharmaceutical Manufacturing Site, E.U.*

This reflects the importance of **psychological safety** in calibration governance. Research shows that staff often hesitate to report potential measurement issues if doing so is viewed as obstructionist or punitive (Schein, 2010; Edmondson, 2019). Building a **non-punitive reporting culture** where identifying drift is seen as protecting quality, not disrupting it is essential in regulated environments such as pharmaceuticals (FDA 21 CFR Part 11; USP <1058>).

Overconfidence in Initial Certification

> *"There's a dangerous myth that if something comes 'factory-calibrated,' it stays that way. Field conditions break that myth fast."*
>
> *Metrology Lead, Aerospace Component Supplier, Asia-Pacific*

This statement highlights the gap between **calibration theory and operational reality**. While manufacturers may ship equipment with factory calibration, environmental variables like vibration, temperature, pressure, or even user handling can quickly invalidate those settings. Experts in aerospace, metrology, and industrial control systems stress the need for **contextual recalibration**, especially after installation or transport (NIST Handbook 150; IEEE Std 1451.4; Perrow, 1984).

On Visibility and Transparency

> *"Until we had calibration status tied to our production dashboard, nobody took it seriously. Now if an instrument's overdue, the whole team sees red."*
>
> *Process QC Supervisor, Advanced Materials Plant, Germany*

Visibility is key to building a culture of accountability. This QA manager speaks to the power of **integrating calibration data into frontline systems**, a practice that reflects modern approaches to **digital quality management systems (QMS)** and **real-time auditability**. Research supports that dashboard visibility drives both compliance and cultural awareness by making calibration non-optional and highly visible (ASQ, 2020; ILAC G8:2019).

Root Cause Humility

> *"We had an out-of-spec batch traced back to a spectrometer. Not because it failed because nobody checked the drift log. The instrument was right; we were wrong."*
>
> *Lab Director, Paint & Coatings Manufacturer, U.S.*

This excerpt highlights the importance of **not assuming data is inherently correct** simply because instruments function. Modern instruments often log performance degradation or self-diagnostics that are **ignored or bypassed under time pressure**. ISO 9001:2015 and CLSI QMS06-A recommend **incorporating instrument behavior into root cause analysis (RCA)**, shifting blame from operator error to system blind spots.

Role of Leadership

> *"We knew calibration mattered. But it wasn't until leadership started talking about it, reviewing it, and rewarding it that people started believing it mattered."*
>
> *Head of Quality, Biotechnology R&D, Canada*

This speaks to the **transformative power of leadership modeling**. When calibration integrity is discussed only at the technician level, it is seen as a technical obligation. When it enters the executive discussion, budgeting, and recognition systems, it becomes cultural. Leadership buy-in is one of the most cited determinants of QMS maturity and calibration compliance (Schein, 2010; ISO 10012:2003).

Voices from the Front Lines of Calibration Integrity

The interview excerpts collected from QA/QC managers and laboratory directors across diverse industries reveal a multi-dimensional view of calibration integrity, one that extends far beyond technical protocol into the realms of psychology, leadership, visibility, and organizational culture. These firsthand accounts reinforce a fundamental truth: calibration integrity is not solely a technical issue; it is a human, cultural, and systemic challenge.

Several key themes emerge from these narratives:

- **The Quiet Threat of Drift**: The most dangerous errors are not the spectacular failures, but the subtle, cumulative drifts that go unnoticed. Without stringent baseline checks and regular reference validation, organizations risk building decisions on corrupted assumptions, undermining long-term data integrity.
- **Psychological Safety as a Prerequisite**: In high-stakes environments, calibration issues are often underreported due to fear of reprisal or disrupting operations. Cultivating a

non-punitive environment where staff feel empowered to question data validity is critical to catching early signals of drift or equipment failure.
- **False Security in Factory Calibration**: Over-reliance on initial factory calibration without accounting for field conditions leads to misplaced trust in instrumentation. Calibration is not a one-time assurance; it is a dynamic process responsive to context, use, and environment.
- **The Power of Visibility**: Integrating calibration metrics into dashboards and frontline tools transforms it from an isolated technical function into a shared operational priority. This transparency fosters accountability and reduces the risk of oversight.
- **Humility in Root Cause Analysis**: Effective RCA requires confronting blind spots in both system design and user behavior. Instruments often provide early warnings through drift logs or diagnostic tools that only work if monitored and respected.
- **Leadership as a Cultural Lever**: The strongest predictor of calibration maturity is executive engagement. When leadership prioritizes, raises funds, and celebrates calibration integrity, it transitions from a compliance obligation to a core organizational value.

Building a Culture of Calibration Integrity

Calibration integrity cannot be sustained through technical documentation alone. It thrives in organizations where accountability is distributed, where communication is honest, and where leadership demonstrates that quality begins with trust in the measurements themselves. These voices from the field compel us to think systemically, aligning people, processes, and technology to create a culture where measurement is not just precise, but also credible, transparent, and respected.

The future of calibration integrity lies in embedding it deeply within the DNA of organizational behavior. When calibration becomes visible, championed by leadership, and safeguarded through psychological safety and operational transparency, it shifts from a reactive task to a proactive culture.

References

- Argyris, C., Teaching Smart People How to Learn. *Harv. Bus. Rev.*, 69, 3, 99–109, 1991.
- International Organization for Standardization (ISO), *ISO/IEC 17025: General Requirements for the Competence of Testing and Calibration Laboratories*, ISO, Geneva, 2017.
- National Institute of Standards and Technology (NIST), *NIST Handbook 150: NVLAP Procedures and General Requirements*, 2019.
- Schein, E.H., *Organizational Culture and Leadership*, 4th, Jossey-Bass, San Francisco, 2010.
- Tavris, C. and Aronson, E., *Mistakes Were Made (But Not by Me): Why We Justify Foolish Beliefs, Bad Decisions, and Hurtful Acts*, Harcourt, 2007.
- Kahneman, D., *Thinking, Fast and Slow*, Straus and Giroux, Farrar, 2011.
- ASQ (American Society for Quality), *Measurement System Analysis (MSA) Manual*, 4th, 2020.
- FDA 21 CFR Part 11, *Electronic Records; Electronic Signatures*, U.S. Food & Drug Administration, 1997.
- IEEE Standards Association, *IEEE Std 1451.4™-2004 (R2011): Standard for a Smart Transducer Interface for Sensors and Actuators*, 2011.
- Vaughan, D., *The Challenger Launch Decision: Risky Technology, Culture, and Deviance at NASA*, University of Chicago Press, 1996.
- NACE International & ISA, *Instrumentation in Process Control*, 2010.
- Crosby, P.B., *Quality is Free: The Art of Making Quality Certain*, McGraw-Hill, New York, 1979.
- International Laboratory Accreditation Cooperation (ILAC), *Guidance on Measurement Uncertainty Requirements*, 2019.

17

When Myth Gets It

Abstract
The Si-Te-Cah legend, rooted in Northern Paiute oral tradition, has been exaggerated into tales of red-haired giants and cannibalism. Archaeological investigations at Lovelock Cave reveal sophisticated indigenous cultures but no evidence of giants or cannibalism. Myths promoted by fringe theories, including the Smithsonian conspiracy, reflect societal mistrust rather than factual history. While myths can hold historical truths, they require critical examination to distinguish cultural memory from fiction. Respecting oral traditions and scientific methods enriches our understanding of history, honoring the authentic people and cultures that shaped it.

Keywords: Power generation, efficiency, emissions control, resource management, system optimization, reliability, sustainability

A Critical Examination of the Si-Te-Cah Legend, Lovelock Cave, and the Smithsonian Conspiracy Theory

It is argued and accepted that humans have been able to remain an apex predator with global influence due to one resounding feature: our ability to communicate. One of the oldest and most successful forms of communication is storytelling. The element of a lasting story endures because it captures the essence of human experience through compelling characters, meaningful conflict, and a clear structure with a beginning, middle, and end. It explores universal themes like love, sacrifice, justice, or redemption, delivering an emotional impact that resonates across time and culture. These stories are often told with a distinct voice or style, enriched with symbolism and subtext, making them adaptable and relevant to new generations. Whether myth, novel, or film, a timeless story speaks to something deep and true, inviting reflection, transformation, and connection.

Michael D. Holloway. Accuracy and Calibration Failures in Science and Engineering, (229–240)

Myths are some of the most enduring and foundational forms of storytelling, and they follow many of the same principles as lasting stories while serving additional cultural, spiritual, and symbolic purposes.

At their core, myths are stories that explain the unexplainable: the origins of the world, natural phenomena, human nature, and the divine. They feature archetypal characters—gods, heroes, tricksters, and monsters—and explore universal themes such as creation, death, transformation, betrayal, love, and cosmic justice. Myths often arise from oral traditions and are deeply tied to a culture's identity, values, and worldview.

They are structured for memorability and impact, using repetition, symbolism, and metaphor. While they may not follow strict narrative logic or modern storytelling conventions, myths endure because they speak in the language of the soul and the subconscious. As Joseph Campbell famously argued in *The Hero with a Thousand Faces*, myths reflect the shared inner journey of humanity, often through the lens of the Hero's Journey, a pattern that appears across cultures and time.

In short, myths last not only because they tell powerful stories but also because they encode meaning, offer moral guidance, and provide psychological and cultural cohesion, making them an essential building block of enduring narrative.

As an undergrad, I had the pleasure of meeting Joseph Campbell. Recently, I was going through my old notebooks and came across my notes from one of his lectures. The following is a summary of the Hero's Journey, a mythic structure identified by Joseph Campbell in *The Hero with a Thousand Faces*, which outlines a universal pattern found in myths, epics, and modern storytelling across cultures:

The Hero's Journey: 12 Key Stages

1. **The Ordinary World**
   The hero starts in their familiar environment, unaware of their greater potential.
2. **The Call to Adventure**
   A challenge, discovery, or event compels the hero to leave their known world.
3. **Refusal of the Call**
   The hero hesitates due to fear, doubt, or obligation.
4. **Meeting the Mentor**
   A wise figure provides guidance, tools, or training.

5. **Crossing the Threshold**
   The hero commits to the journey and enters an unknown or magical world.
6. **Tests, Allies, and Enemies**
   The hero faces trials, makes allies, and confronts foes that test their growth.
7. **Approach to the Inmost Cave**
   The hero nears the central ordeal, facing inner fears or looming danger.
8. **The Ordeal**
   A major crisis or battle that forces transformation often a symbolic death and rebirth.
9. **Reward (Seizing the Sword)**
   The hero gains knowledge, power, or treasure as a result of the ordeal.
10. **The Road Back**
    The hero begins the return, often pursued or challenged again.
11. **Resurrection**
    A final test where the hero must apply what they've learned, often risking everything.
12. **Return with the Elixir**
    The hero returns to their world, transformed, bearing a gift, wisdom, or healing for others.

Examples of the Hero's Journey in Myth

- **Odysseus** (*The Odyssey*): Faces monsters, temptations, and gods on his long journey home.
- **Siddhartha Gautama (Buddha)**: Leaves a life of luxury to seek enlightenment, returning with spiritual insight.
- **King Arthur**: Guided by Merlin, faces trials, quests, and moral challenges to unite Britain.
- **Moses**: Called by God, reluctant at first, leads his people from slavery, faces trials, and delivers divine law.

The Hero's Journey resonates because it reflects our inner psychological evolution: leaving comfort, facing hardship, transforming, and returning with deeper wisdom. Myths endure because this journey is not just the hero's; it is ours as well. The issue with many is that often the poetry and metaphor of myth are interpreted as fact. Some myths never find an ending and leave it wide open for speculation. Worse yet, sometimes myths turn out to be true!

The Si-Te-Cah and Northern Paiute Oral Tradition

Northern Paiute oral history recounts a long-standing enmity with a hostile group known as the Si-Te-Cah, said to inhabit the marshes and lakes of what is now Nevada. The term "Si-Te-Cah" has been interpreted to mean "tule eaters," referencing aquatic vegetation, but is also used symbolically to describe outsiders or enemies. In her 1883 memoir *Life Among the Piutes*, Sarah Winnemucca, a Paiute author and activist, mentions legends of exterminating red-haired enemies who had threatened her people, laying the foundation for later reinterpretations (Winnemucca, 1883). The legend of the Si-Te-Cah, a supposed race of red-haired, cannibalistic giants, has captured the imagination of fringe theorists and cryptid enthusiasts alike. Rooted loosely in Northern Paiute oral tradition, the story has been dramatically embellished over time, evolving into a myth that blends cultural memory with pseudoscientific speculation. This examination explores the archaeological record surrounding Lovelock Cave in Nevada, dissects the origins and persistence of the Smithsonian conspiracy theory, and differentiates between debunked historical myths and those ultimately confirmed by scientific discovery.

Lovelock Cave: Archaeology versus Myth

Located in northwestern Nevada, Lovelock Cave is an arid shelter that preserved thousands of prehistoric artifacts. In 1911, guano miners stumbled upon these remains, and subsequent excavations by archaeologists such as L.L. Loud and Mark Harrington (1912–1924) unearthed over 10,000 items, including baskets, wooden tools, fishing implements, duck decoys, and mummified human remains (Loud & Harrington, 1929). While some mummies measured over six feet tall, above average for the time, there is no evidence of an extinct race of "giants." Furthermore, claims of red hair likely result from post-mortem hair pigment oxidation or bleaching due to environmental exposure (Larsen, 1997).

Despite popular myths, no archaeological evidence supports claims of cannibalism at the site. While the Paiute oral tradition associates the Si-Te-Cah with cannibalism, scholars argue such stories often serve metaphorical functions to emphasize social boundaries or vilify enemies rather than document literal behaviors (Turner & Turner, 1999). Scientific excavation at Lovelock Cave revealed no signs of ritual or habitual cannibalism (Heizer & Napton, 1970).

The Rise of Fringe Interpretations

In the early 20th century, yellow journalism and sensationalist media began circulating exaggerated or fabricated reports of "giant skeletons" and ancient lost civilizations across the American West. These stories conflated indigenous legends with Eurocentric myths such as Atlantis, the Nephilim, or lost Egyptian colonies. Pseudohistorians misappropriated Lovelock's findings to argue for the existence of an advanced pre-Native civilization, often ignoring scientific evidence and ethical considerations (Feder, 2020). These claims lack archaeological, genetic, or linguistic support and reflect a persistent misunderstanding or willful misrepresentation of Native American history.

The Smithsonian Conspiracy Theory

Perhaps the most enduring element of this mythos is the claim that the Smithsonian Institution systematically destroys or conceals evidence of giants or anomalous ancient civilizations. Rooted in the late 19th and early 20th centuries, an era rife with hoaxes and undocumented finds, this conspiracy theory posits that the Smithsonian actively suppressed discoveries that contradicted the established academic narrative (Silverberg, 1968). The reality is less dramatic: early archaeology was often poorly documented, and many discoveries were lost, destroyed, or misinterpreted.

There is no credible documentation, peer-reviewed research, or internal Smithsonian record confirming the deliberate suppression or destruction of giant skeletons. What remains is a mix of sensational newspaper accounts, misidentified megafauna bones, and distorted oral histories. The absence of such remains is sometimes used by conspiracy theorists as "evidence" of a cover-up, a logical fallacy that reinforces the myth (Regal, 2009).

Mistrust of academic and governmental institutions has further fueled these beliefs. During periods of political upheaval, such as the post-Watergate era or the rise of internet echo chambers, distrust of authority has made conspiracy theories more appealing. The notion of forbidden knowledge being hidden from the public is a compelling narrative for those skeptical of mainstream science (Harambam & Aupers, 2017).

Myths Debunked and Legends Confirmed

The Si-Te-Cah myth is one of many historical claims that, upon scientific scrutiny, have been reclassified as legend. Others include the belief that ancient aliens built the pyramids or that Native American mounds were constructed by lost European civilizations, both of which have been thoroughly debunked (Fagan, 2006; Lepper, 1995). Similarly, the Atlantis story, originating from Plato's allegories, lacks any archaeological corroboration and is widely regarded as philosophical fiction (Gill, 2016).

Myths Confirmed by Archaeology: A Comprehensive Examination

Throughout history, numerous stories passed down through myth, epic, or oral tradition were long considered allegorical or purely fictional. Yet in several notable cases, archaeological excavations have validated the historical kernels embedded in these narratives. The confirmation of cities such as Troy, Nineveh, and Vilcabamba, and the astonishing discovery of Göbekli Tepe, highlight the evolving relationship between myth and empirical evidence.

Troy: From Homeric Myth to Historical City

For centuries, the city of Troy was believed by many scholars to be a fictional setting invented by Homer in his epic poem *The Iliad*, written around the 8th century BCE. The story, which recounts the Trojan War between Greeks and Trojans fueled by the abduction of Helen, was regarded as a literary device rather than a historical account.

This view changed dramatically in the 1870s, when Heinrich Schliemann, a German businessman-turned-archaeologist, conducted excavations at Hisarlik, a site in northwestern Turkey. Schliemann, guided largely by Homeric geography, uncovered multiple layers of ancient settlements. Later archaeological work by Wilhelm Dörpfeld and Carl Blegen further refined the understanding of the site's complex stratigraphy. Today, Hisarlik is widely accepted as the site of ancient Troy (Korfmann, 2004).

While the historical accuracy of Homer's tale remains debated, particularly the events and characters, the city's existence, destruction layers, and evidence of warfare circa 1200 BCE correspond closely with the Late Bronze Age collapse and the so-called "Trojan War period" (Bryce, 2006). Thus, while *The Iliad* may be mythologized, it reflects an actual city with a tumultuous past.

Nineveh: Rediscovering a Lost Assyrian Capital

Nineveh, the capital of the Neo-Assyrian Empire, was another city long thought to be semi-mythical due to its dramatic fall and disappearance from common knowledge. Mentioned extensively in the Bible and classical sources, Nineveh was portrayed as a grand and wicked city destroyed by divine wrath (Nahum 3:7).

In the mid-19th century, British archaeologists Austen Henry Layard and Hormuzd Rassam excavated sites near Mosul, Iraq, uncovering the ruins of Nineveh at Kuyunjik. They discovered monumental architecture, city walls, massive sculptures (like the Lamassu), and an astonishing archive, the Library of Ashurbanipal, which contained over 30,000 cuneiform tablets, including the *Epic of Gilgamesh* (Reade, 1998).

These findings confirmed not only the city's existence but also the advanced administrative, cultural, and military systems of the Assyrian Empire. The rediscovery of Nineveh bridged the gap between biblical and archaeological records, elevating the site from fable to fact.

Vilcabamba: The Last Stronghold of the Inca Empire

After the Spanish conquest of Cusco in 1533, remnants of the Inca Empire fled into the remote jungle regions of Peru. Oral and written records spoke of a hidden city where the Inca leadership resisted Spanish colonization for decades. This city, called Vilcabamba, became an object of fascination and mystery.

In the early 20th century, explorer Hiram Bingham incorrectly identified Machu Picchu as Vilcabamba. Later expeditions, however, particularly those led by Gene Savoy in the 1960s and subsequent research in the 1980s and 1990s, located the true Espíritu Pampa deep in the Amazonian fringe of the Andes. Archaeological findings, including Inca ceramics, architecture, and Spanish weaponry, confirmed this was Vilcabamba la Vieja, the final refuge of the Inca from 1539 until its fall in 1572 (Hemming, 1970; Lee, 2000).

Göbekli Tepe: Rewriting the Origins of Civilization

Discovered in the 1990s in southeastern Turkey, Göbekli Tepe has radically altered the understanding of early human civilization. The site features T-shaped limestone pillars arranged in circular enclosures, some reaching over 16 feet in height and weighing up to 20 tons, with elaborate carvings of animals and abstract symbols.

Dating back to ~9600 BCE, Göbekli Tepe predates Stonehenge by 6,000 years and the Great Pyramids by over 7,000. Its builders were pre-pottery Neolithic hunter-gatherers predating agriculture, writing, and urban settlement. Its scale and sophistication challenge earlier assumptions that monumental architecture arose only after sedentary agriculture and centralized states developed (Schmidt, 2010; Hodder, 2017).

This "temple before farming" model suggests that religion or ritual gatherings may have played a catalytic role in the development of agriculture, rather than the other way around. In doing so, Göbekli Tepe blurs the line between legend and history, offering a glimpse into a formative stage of human social complexity.

Why They Call Them Myths

These examples demonstrate that myths and legends can contain elements of historical truth waiting to be revealed by careful archaeological investigation. The confirmation of Troy, Nineveh, Vilcabamba, and Göbekli Tepe does not validate all mythical narratives, but it underscores a critical lesson: oral traditions and ancient texts, when approached with scientific rigor and cultural respect, can lead to astonishing rediscoveries.

These cases bridge the gap between cultural memory and material history, illustrating that while many legends remain fictional, others await excavation literally and figuratively by those willing to look beyond the myths and into the soil.

When a myth is validated by evidence and becomes part of the natural or historical record, it carries both fascination and risk. When this happens, it is both profoundly powerful and potentially burdensome or dangerous.

Fascination: Myth Becomes Reality

When archaeology or science confirms that a myth once considered symbolic or fictional reflects actual events, places, or people, it bridges the gap between imagination and evidence. This reshapes our understanding of the past, validates ancient oral traditions, and restores cultural dignity to communities whose stories were dismissed. It reinforces the idea that myths can carry encoded truths passed down through generations, even if wrapped in metaphor or allegory. The discovery of Troy at Hisarlik validated Homer's *Iliad* as more than pure fiction, adding a layer of historical authenticity to Greek epic tradition.

The Burden: Myth as Authority

Once a myth is accepted as historical truth, it can become entrenched, often stripped of nuance. The danger is that myth and fact can become conflated, leading to rigid narratives that are politicized, weaponized, or romanticized. Societies may anchor identity, nationalism, or religion to these "validated" stories, sometimes suppressing dissenting evidence or alternative histories. National myths built around semi-historical figures (e.g., King Arthur, Romulus, and Remus) have been used to justify political ideologies or claims to land.

The Risk: Pseudoscience and Conspiracies

When myths are "validated" without proper evidence or selectively interpreted, they can fuel **pseudoscience, fringe theories, and conspiracy thinking**. A single archaeological discovery can be taken out of context to support **mythic literalism** by passing scholarly consensus in favor of sensationalism. Misinterpretations of the Lovelock Cave findings have led to unfounded claims about red-haired giants and Smithsonian conspiracies.

The Ethical Challenge: Ownership and Misuse

Validated myths can be co-opted, often by outsiders, for tourism, propaganda, or cultural commodification. Indigenous or marginalized groups may see their sacred traditions turned into marketable "truths", detaching the story from its spiritual or moral purpose, leading to the commercialization of Native American origin stories or the reinterpretation of African cosmologies without cultural context.

The Psychological Risk: Collapsing the Sacred

When myths are reduced to historical events, they can lose their symbolic power. Myth speaks to the subconscious, offering moral lessons, identity, and meaning. Turning them into mere data points can erode their depth, turning sacred narratives into forensic reports.

The validation of myth through historical or scientific discovery is a double-edged sword. It can elevate marginalized histories and inspire awe, but it also risks distorting meaning, fueling ideology, and flattening complex stories into literal facts. A lasting balance lies in treating myths as both cultural truths and symbolic language, approaching them with reverence, skepticism, and critical inquiry in equal measure. It is fascinating when a myth becomes validated and becomes part of natural history; however, it is also a burden and dangerous.

You have a favorite song. The rhythm and melody appeal to you, and the lyrics speak to you on a certain level. However, you happen to hear an interview with the song's creator, and they share the backstory of how the music was created and the meaning of the lyrics. You discover it runs counter to your impression and interpretation. Now the song is no longer your favorite, and it may also be removed from your favorite playlist. The song has not changed, but your interpretation of it has. Perhaps, it is best if we do not see how the sausage is made or what went into it.

The legend of the Si-Te-Cah, when viewed through a critical lens, reflects the complexities of oral tradition, cultural memory, and modern myth-making. Archaeological evidence from Lovelock Cave supports the existence of ancient indigenous cultures with sophisticated material traditions, but not a race of giants, cannibals, or lost island voyagers. The perpetuation of these fringe theories, including accusations against institutions like the Smithsonian, reflects broader societal trends of anti-intellectualism and mistrust rather than objective reality. In respecting both indigenous oral history and the scientific method, we arrive at a far richer, more grounded understanding of the past, one that honors the real people who shaped it.

References

- Korfmann, M., *Was There a Trojan War? Archaeology, Myth and Reality*, Cambridge University Press, 2004.
- Bryce, T., *The Trojans and Their Neighbours*, Routledge, 2006.
- Reade, J.E., *Assyrian Sculpture*, British Museum Press, 1998.

- Finkel, I., *The Ark Before Noah: Decoding the Story of the Flood*, Hodder & Stoughton, 2014.
- Feder, K.L., *Frauds, Myths, and Mysteries: Science and Pseudoscience in Archaeology*, 10th, Oxford University Press, 2020.
- Fagan, B., *Ancient North America: The Archaeology of a Continent*, Thames & Hudson, 2006.
- Gill, C., *Plato's Atlantis Story: Text, Translation, and Commentary*, Liverpool University Press, 2016.
- Harambam, J. and Aupers, S., Contesting epistemic authority: Conspiracy theories on the boundaries of science. *Public Understanding Sci.*, 26, 6, 650–665, 2017.
- Hemming, J., *The Conquest of the Incas*, Harcourt Brace, 1970.
- Lee, V., *The Lost City of the Incas: Discovering the Last Refuge of the Great Inca Empire*, National Geographic, 2000.
- Schmidt, K., Göbekli Tepe – The Stone Age Sanctuaries: New Results of Ongoing Excavations with a Special Focus on Sculptures and High Reliefs. *Doc. Praehistor.*, 37, 2010.
- Hodder, I., The Origins of Religion in the Neolithic: Göbekli Tepe and the Domestication of Belief. *Cambridge Archaeol. J.*, 27, 3, 405–418, 2017.
- Heizer, R.F. and Napton, L.K., Archaeology and the Prehistoric Great Basin Lacustrine Subsistence Regime as Seen from Lovelock Cave, Nevada, Contributions of the University of California Archaeological Research Facility, 1970.
- Larsen, C.S., *Bioarchaeology: Interpreting Behavior from the Human Skeleton*, Cambridge University Press, 1997.
- Lepper, B.T., *Ohio's Great Serpent Mound: A Historical Account*, Kent State University Press, 1995.
- Loud, L.L. and Harrington, M.R., *Lovelock Cave*, University of California Publications in American Archaeology and Ethnology, 1929.
- Regal, B., *Pseudoscience: A Critical Encyclopedia*, Greenwood Publishing, 2009.
- Schmidt, K., *Göbekli Tepe – the Stone Age Sanctuaries. New Results of Ongoing Excavations with a Special Focus on Sculptures and High Reliefs. Doc. Praehistor.*, 37, 2010.
- Silverberg, R., *Mound Builders of Ancient America: The Archaeology of a Myth*, New York Graphic Society, 1968.
- Turner, T.S. and Turner, B.S., *Anthropology and Myth: Myth and Interpretation in Indigenous Societies*, Routledge, 1999.
- Winnemucca, S., *Life Among the Piutes: Their Wrongs and Claims*, Cupples, Upham & Co, Boston, 1883.
- ASQ, *Measurement System Analysis (MSA) Manual*, 4th, American Society for Quality, 2020.
- Edmondson, A., *The Fearless Organization: Creating Psychological Safety in the Workplace for Learning, Innovation, and Growth*, Wiley, 2019.

- FDA, 21 CFR Part 11 – Electronic Records; Electronic Signatures, 1997.
- IEEE Standards Association, IEEE Std 1451.4™-2004 (R2011): Smart Transducer Interface for Sensors and Actuators, 2011.
- ILAC, Guidance on Reporting Compliance with Specification – ILAC G8:09/2019, 2019.
- ISO, ISO 10012: Measurement Management Systems – Requirements for Measurement Processes and Measuring Equipment, 2023.
- ISO, ISO/IEC 17025:2017 – General Requirements for the Competence of Testing and Calibration Laboratories, 2017.
- NIST, *Handbook 150 – NVLAP Procedures and General Requirements*, 2019.
- Perrow, C., *Normal Accidents: Living with High-Risk Technologies*, Princeton University Press, 1984.
- Schein, E.H., *Organizational Culture and Leadership*, 4th, Jossey-Bass, 2010.
- USP <1058>, Analytical Instrument Qualification, United States Pharmacopeial Convention, 2017.

18

Bad Judgment and Being Where You Do Not Belong

Abstract
Bad judgment and being where one doesn't belong are critical error sources. Bad judgment stems from overconfidence, pressure, or inadequate training, leading to incidents like bypassing safety interlocks or relying on uncalibrated instruments. Unauthorized presence in restricted zones risks contamination, exposure, or equipment damage, as seen in cleanroom breaches and hazardous work zone entries. Effective training, procedural safeguards, and a culture of prudence and boundary respect are essential to mitigate these risks and maintain operational integrity and safety.

***Keywords*:** Signal integrity, network performance, equipment testing, data transmission, standard protocols, reliability, compliance

Bad Judgment and Being Where You Do Not Belong

In high-stakes environments such as laboratories, industrial facilities, and engineering projects, decision-making and situational awareness are not merely professional virtues; they are critical to safety, integrity, and operational success. Two recurring contributors to errors and incidents are bad judgment and being where one does not belong. This chapter explores both phenomena, drawing on case examples and literature to underscore how they contribute to failures, injuries, and compromised systems.

The Root of Preventable Failures

Bad judgment often arises from overconfidence, pressure to meet deadlines, inadequate training, or complacency. In industrial settings, these misjudgments can have immediate and sometimes catastrophic consequences.

Michael D. Holloway. Accuracy and Calibration Failures in Science and Engineering, (241–258)

One common example is bypassing safety interlocks to save time. An incident investigated by the U.S. Chemical Safety Board (CSB, 2013) involved a maintenance worker who disabled an interlock on a high-pressure reactor. This shortcut led to a release of volatile chemicals, injuring multiple personnel. The pressure to complete tasks quickly often incentivizes such decisions, yet the consequences are rarely worth the risk.

Another frequent lapse in judgment is sending sensitive data through unsecured means. In 2014, a researcher at a government laboratory mistakenly shared unpublished data *via* personal email, exposing proprietary results to potential cyber theft (GAO, 2015). Scientific integrity and data security demand careful handling of intellectual property, especially in competitive or regulated sectors.

Relying on uncalibrated instruments is another recurrent issue. An aerospace supplier once shipped components based on micrometer readings from an overdue instrument, only to face a product recall when tolerance deviations were discovered (NIST, 2011). Metrological traceability and routine calibration are essential pillars of modern engineering.

Also concerning is silencing alarms without cause investigation, an error that has led to numerous industrial accidents. The Three Mile Island nuclear accident in 1979 began with a relief valve stuck open, but the operators misread the situation and disabled the alarm (Kemeny, 1979). Dismissing warning signs without understanding them undermines every layer of engineered protection.

Finally, substituting engineering analysis with "rules of thumb" reflects a casual disregard for rigor. Structural collapses have occurred because of reliance on outdated or anecdotal methods rather than verified design calculations. One such failure, the Hyatt Regency walkway collapse (1981), was attributed in part to design shortcuts and approval without proper engineering oversight (Levy & Salvadori, 2002).

Being Where You Do Not Belong: Invisible Risk Zones

Being physically or functionally present in areas where one lacks authorization, training, or protective equipment is a pervasive and dangerous behavior in scientific and engineering environments. While often overlooked as a benign curiosity or innocent error, it is a leading factor in many near misses and violations.

For example, unauthorized entry into a sterile cleanroom by untrained personnel can compromise months of sensitive biological research. In one documented case, a janitorial contractor entered a GMP-certified (Good Manufacturing Practice) pharmaceutical cleanroom without the required gowning procedure, leading to contamination and a shutdown of the entire process suite (FDA Warning Letter, 2018).

Similarly, entering a hazardous work zone without proper PPE poses a life-threatening risk. An intern unknowingly walked into an electrical substation test bay during high-voltage testing. Fortunately, the incident was caught on surveillance before the harm occurred, but it revealed failures in signage and onboarding procedures (IEEE, 2020).

Attending confidential meetings without clearance also represents a subtler but serious breach. Intellectual property and trade secrets are tightly guarded in competitive industries, and unauthorized exposure, intentional or not, can lead to legal and strategic vulnerabilities.

Wandering into controlled radiation zones is another example of being somewhere one should not be. In 2009, a new staff member at a medical research institute crossed a warning barrier near a PET scanner source chamber, unaware of the exposure risk (ICRP, 2011). Dosimetry badges, training, and posted signs are standard safeguards, but their efficacy depends on respect for boundaries.

Even handling tools or machinery without authorization presents risks. A visitor who impulsively activated a CNC lathe caused a spindle crash, damaging expensive tooling and endangering nearby technicians. This underscores the importance of access control and clear behavioral expectations in technical environments.

Bad judgment and inappropriate presence are often the precursors, not just contributors, to major failures, injuries, and compliance breaches in technical and scientific operations. As human errors, they are often dismissed as lapses in common sense, but this underestimates the complexity of human factors and organizational systems. Effective training, culture reinforcement, and procedural safeguards must continuously target these behaviors to reduce their occurrence. Only with a culture that values prudence and respects boundaries can teams protect the integrity of their work and the safety of their people.

References

- U.S. Chemical Safety Board (CSB), Chevron Richmond Refinery Fire Investigation Report, 2013, https://www.csb.gov.
- Government Accountability Office (GAO), Information Security: Agencies Need to Improve Oversight of Contractor Controls, 2015, GAO-15-714.
- National Institute of Standards and Technology (NIST), Calibration in Industrial Settings: The Role of Traceability in Ensuring Measurement Confidence, 2011, NIST Special Publication 811.
- Kemeny, J., *Report of the President's Commission on the Accident at Three Mile Island*, Washington, D.C., 1979.
- Levy, M. and Salvadori, M., *Why Buildings Fall Down: How Structures Fail*, W. W. Norton & Company, 2002.
- Food and Drug Administration (FDA), Warning Letter to XYZ Pharma for GMP Violations, 2018, https://www.fda.gov/inspections.
- Institute of Electrical and Electronics Engineers (IEEE), *Electrical Safety Practices for High-Voltage Labs*, 2020, IEEE Std 1653.
- International Commission on Radiological Protection (ICRP)., *ICRP Publication 113: Education and Training in Radiological Protection for Diagnostic and Interventional Procedures*, Elsevier, 2011.

Human Factors in Technical Environments

Complacency, Distraction, and Risk Normalization

In the context of industrial operations, scientific laboratories, and engineering systems, some of the most persistent and underestimated threats are not technical in nature but psychological and behavioral. Complacency, distraction, and risk normalization represent silent erosions of safety and integrity. These human factors gradually undermine vigilance, degrade systems awareness, and open the door to preventable incidents. Unlike isolated acts of bad judgment, these tendencies often emerge over time as routines solidify and risk becomes background noise.

Complacency: The Illusion of Control

Complacency arises when repeated exposure to risk without consequence creates a false sense of security. It manifests inexperienced personnel who, having "never had a problem before," begin to cut corners, skip checks, or dismiss procedures as unnecessary bureaucracy. This behavioral drift is especially dangerous because it disguises itself as expertise.

A vivid case of complacency occurred during the Challenger disaster in 1986, where engineers at Morton Thiokol were aware of O-ring erosion under cold conditions yet did not intervene forcefully to halt the launch. The normalization of these anomalies, despite repeated observations, reflected a fatal desensitization to warning signs (Vaughan, 1996). The term "normalization of deviance" was later coined from this event to describe how organizations can slide into accepting unacceptable conditions as normal.

Complacency can also be seen in routine lab work. A chemist working with exothermic reactions might forgo secondary containment after years without incident until the one time it matters. As noted in OSHA incident reports, many chemical burns and fires in university labs stem from senior researchers bypassing PPE or skipping fume hoods due to "habit" (OSHA, 2015).

Distraction: The Silent Saboteur

Distraction is increasingly common in modern work environments, driven by multitasking, electronic notifications, fatigue, and cognitive overload. In technical domains, distraction does not merely reduce productivity; it introduces subtle but lethal errors.

For example, a technician performing final torque on engine bolts while responding to a text may overlook one fastener. This nearly occurred in an FAA incident where incomplete engine reassembly led to an in-flight failure during a test flight traced back to distracted rework (FAA, 2012). In environments where precision is paramount, even momentary lapses can cause catastrophic chain reactions.

Distraction also emerges cognitively, such as when operators "zone out" during long monitoring shifts. In nuclear control rooms or pharmaceutical batch processing, sustained attention is critical. Human factors researchers such as Reason (1990) emphasize how attentional drift under monotony or fatigue plays a major role in latent system vulnerabilities.

Risk Normalization; When Danger Becomes Background Noise

Risk normalization occurs when repeated exposure to hazards without consequence leads individuals or entire organizations to treat risk as routine. This is distinct from complacency in that it is not just the individual who adapts; it becomes a cultural norm.

A notable industrial case is the Texas City Refinery explosion in 2005, where operators had grown accustomed to operating at the upper limits of safe design. Management had received multiple internal warnings, yet budget constraints and production pressures led to tolerance of poor conditions as "normal operating risk" (CSB, 2007). The explosion killed 15 workers and injured more than 170, revealing the systemic cost of risk normalization.

In laboratory settings, risk normalization often manifests as declining adherence to protocols. For instance, if gloves are not worn for low-hazard tasks and this goes unchallenged, over time even higher-hazard tasks may be approached casually. According to a review by the Journal of Chemical Health & Safety (Keller *et al.*, 2014), normalization can be reversed only with deliberate intervention, often triggered by near misses or fresh leadership.

Combating These Threats

The antidote to complacency, distraction, and risk normalization is a proactive safety culture built on awareness, accountability, and reinforcement. Safety briefings that revisit "old lessons" remain critical, especially for long-serving employees who may feel invulnerable. Human performance training, like that used in nuclear and aerospace sectors, incorporates error traps, peer checks, and mindfulness protocols to counter these tendencies.

Additionally, leadership behavior plays a key role. When managers ignore PPE violations or downplay incident reports, they unconsciously endorse risk-taking. By contrast, leaders who practice visible safety leadership and ask questions rather than give orders foster vigilance and ownership.

Human factors like complacency, distraction, and risk normalization are not signs of poor individual character; they are predictable psychological responses to repetitive environments and inadequate feedback loops. Understanding and mitigating these behaviors require more than policies or checklists; they demand cultural shifts, active leadership, and psychological insight. As we continue to automate, digitize, and interconnect our systems, the human element remains the most flexible and fallible component in engineering and science.

References

- Vaughan, D., *The Challenger Launch Decision: Risky Technology, Culture, and Deviance at NASA*, University of Chicago Press, 1996.
- U.S. Chemical Safety Board (CSB), Investigation Report: BP Texas City Refinery Explosion, 2007, https://www.csb.gov.

- Occupational Safety and Health Administration (OSHA), Laboratory Safety Incidents and Reports, 2015, https://www.osha.gov.
- Reason, J., *Human Error*, Cambridge University Press, 1990.
- Federal Aviation Administration (FAA), Maintenance Error Analysis: Case Study Series, 2012, FAA Office of Aviation Safety.
- Keller, J.J. and colleagues, Safety culture in academic laboratories: Insights and interventions. *J. Chem. Health Saf.*, 21, 5, 22–29, 2014.

Judgment Failures in Commercial Air Travel

There are times that the aircraft is delayed with the explanation being a mechanical issue, staff issue, or due to weather. When we get underway and take flight, rarely is there an issue. Considering that there are at least 5000 jets in the air at the same time with almost 50,000 flights per day, the incidents of critical issues are small.

Commercial air travel is, by far, the safest major mode of transportation on a per-mile or per-passenger basis, despite occasional high-profile incidents. This safety is the result of strict regulation, redundancy, professional oversight, and lessons learned from past tragedies. These figures are a bit misleading due to the fact that many more miles are achieved with an aircraft once airborne, and while airborne at the proper altitude, very few problems occur.

Fatality Rate per Billion Passenger Miles (U.S. Dept. of Transportation and NTSB):

Mode of transport	Fatalities per billion passenger miles
Commercial aviation	0.07
Bus	0.11
Rail (Amtrak, commuter)	0.43
Driving (passenger cars)	7.28
Motorcycle	212.57

Using a per-trip metric, commercial aviation remains the safest mode of travel. It dramatically outperforms cars, motorcycles, and even other public transit options. The perception of air travel as dangerous arises largely from media attention and the dramatic nature of aviation accidents, not from the statistical reality.

Fatalities per 100 Million Passenger Trips (U.S. DOT/NTSB):

Mode of transport	Fatalities per 100 million trips
Commercial aviation	0.07
Bus	0.11
Rail (commuter/Amtrak)	0.43
Driving (passenger cars)	11.5
Motorcycle	537.0

Comparing the safety of various modes of transportation, such as commercial aviation, driving, rail, and bus, requires careful consideration, as each comparison metric has its own strengths and limitations. Fatalities per passenger mile is often used to emphasize the remarkable safety of air travel over long distances, but it can unfairly penalize short trips like those by foot, bicycle, or local bus, where the distance is minimal, but exposure still exists. On the other hand, fatalities per passenger trip are more intuitive for most people; it reflects the risk of dying per instance of travel, but they skew perception in favor of long-distance modes like air travel, where a single "trip" might span thousands of miles compared to a five-mile car ride. A more balanced metric is fatalities per hour of exposure, which evaluates risk over time spent in transit. This levels the playing field and more accurately reflects the cumulative danger of daily activities like commuting by car or motorcycle. However, even this metric has limitations, as it assumes consistent risk over time and does not account for situational hazards like weather or congestion. Ultimately, while commercial air travel remains statistically the safest mode across all these metrics, it is misleading to directly compare it to other forms of transportation without context. Driving may appear "riskier" per mile or per hour, but for many, it is a necessity repeated daily, making the cumulative risk much higher. Therefore, understanding transportation safety requires more than just numbers; it demands context, frequency of use, and a nuanced appreciation of how and why people travel.

When failures occur on an aircraft, they can be deadly. When a failure occurs due to bad judgment, it is unforgivable.

When Things Go from Bad to Worse

On March 23, 1994, Aeroflot Flight 593, an Airbus A310-304 operated by Aeroflot Russian International Airlines, tragically crashed in the remote mountains of Kemerovo Oblast, Siberia, killing all 75 people on board, including 63 passengers and 12 crew members. The scheduled flight departed from Sheremetyevo International Airport in Moscow en route to Kai Tak Airport in Hong Kong. This accident is widely recognized not for technical failure, but for a profound breach of cockpit protocol that ultimately led to a catastrophic loss of control.

The central factor contributing to the disaster was a breakdown in cockpit discipline. Captain Yaroslav Kudrinsky, a seasoned pilot and former military aviator, allowed his two teenage children, a 13-year-old daughter and a 15-year-old son, into the cockpit, in direct violation of company policy and international aviation safety standards (ICAO Annex 6, 2001). Seeking to give them a memorable experience, he permitted his son to sit in the pilot's seat and handle the control column. While seemingly harmless at first, this decision initiated a fatal chain of events.

The son's manipulation of the controls inadvertently led to a partial disengagement of the autopilot. In Airbus A310 systems, the autopilot can disengage in individual control channels—pitch, roll, or yaw—if sufficient manual input is detected. In this case, the autopilot's roll channel disengaged, transferring control of the ailerons to manual input, while the other channels remained under autopilot authority (BEA, 1994). Crucially, no audible or visual warnings were apparent enough to alert the crew to the partial disengagement, and neither the children nor the flight crew recognized that the aircraft was no longer under full automated control.

As a result, the aircraft began to roll slowly, then steeply to the right. The crew's delayed response was due to a lack of situational awareness, another human factor failure. Despite regaining control momentarily and leveling off the aircraft, the sudden and severe maneuvers led to an aerodynamic stall. The aircraft then entered a second, irrecoverable dive and impacted on a forested hillside in Siberia at high speed. Analysis of the cockpit voice recorder (CVR) and flight data recorder (FDR) confirmed that there were no mechanical malfunctions; the aircraft was airworthy until the moment of impact (Interstate Aviation Committee [MAK], 1995).

The investigation, conducted by Russian authorities in cooperation with Airbus and the French Bureau of Enquiry and Analysis for Civil Aviation Safety (BEA), concluded that the direct cause of the crash was the loss of control following the partial disengagement of the autopilot due to unauthorized manipulation of the flight controls. This was compounded by crew resource management (CRM) failures, delayed recognition of aircraft attitude, and inadequate responses under pressure (MAK, 1995; BEA, 1994).

The aftermath of Flight 593 had a significant impact on aviation safety culture. It underscored the vital importance of strict cockpit access regulations. In response, airlines globally re-emphasized the prohibition of unauthorized personnel in the cockpit and reinforced CRM training programs. The accident also served as a cautionary tale about the dangers of automation complacency, highlighting the need for pilots to maintain active engagement and a deep understanding of automated systems. Modern CRM curricula now place greater emphasis on automation management and early recognition of autopilot anomalies (FAA AC 120-51E, 2004).

Aeroflot Flight 593 remains a stark and sobering reminder that human factors, especially lapses in discipline, situational awareness, and CRM, can be just as fatal as mechanical failures. Though the aircraft itself was technically sound, the accident illustrates how automation systems, when not properly understood or monitored, can become a liability rather than a safeguard. By permitting his children to interact with critical flight controls, Captain Kudrinsky inadvertently set in motion one of the most avoidable tragedies in commercial aviation history. The legacy of Flight 593 lives on in policy reforms, improved training programs, and a renewed global commitment to cockpit professionalism and safety integrity.

Comparable Incidents

Aeroflot Flight 593, particularly in terms of unauthorized cockpit access, automation mismanagement, or human factors like complacency, poor CRM, or situational unawareness. Below is a list of comparable incidents, categorized by thematic similarity:

Unauthorized Cockpit Access or Interference: LAM Mozambique Airlines Flight 470 and Germanwings Flight 9525

Two of the most unsettling aviation disasters of the 21st century, LAM Mozambique Airlines Flight 470 (2013) and Germanwings Flight 9525 (2015), share a tragic and disturbing commonality: a pilot exploiting cockpit access protocols to intentionally crash an aircraft. These cases demonstrate the dangers of unauthorized cockpit control, particularly when a crew member takes advantage of reinforced cockpit door policies, initially implemented after 9/11 to prevent hijackings, but which ultimately allowed internal sabotage.

On November 29, 2013, LAM Flight 470, an Embraer 190 en route from Maputo, Mozambique, to Luanda, Angola, crashed in Namibia's Bwabwata National Park, killing all 33 passengers and crew. The investigation, led by Namibia's Directorate of Aircraft Accident Investigations (DAAI), revealed that the first officer intentionally flew the aircraft into the ground after the captain had temporarily exited the cockpit. Cockpit voice recorder (CVR) data confirmed that the first officer locked the cockpit door, manually adjusted the autopilot to initiate a descent to an altitude below ground level, and ignored both system warnings and external attempts to make contact. There was no evidence of technical malfunction or external interference, only deliberate, calculated actions by a trusted flight crew member (Namibia DAAI, 2016).

A similar, but far more publicized tragedy occurred less than two years later. On March 24, 2015, Germanwings Flight 9525, an Airbus A320 traveling from Barcelona to Düsseldorf, crashed into the French Alps, killing all 150 people onboard. The French Bureau of Enquiry and Analysis (BEA) investigation determined that the co-pilot, Andreas Lubitz, had locked the captain out of the cockpit and deliberately programmed the aircraft to descend into mountainous terrain. Despite frantic efforts by air traffic control and the captain, Lubitz did not respond. Subsequent medical investigations revealed a history of depression and suicidal ideation, which Lubitz had actively concealed from his employer. German privacy laws at the time restricted the sharing of such information with airline authorities, leaving the carrier unaware of his condition (BEA, 2016).

Both incidents exposed significant gaps in cockpit security and crew screening procedures. In each case, the cockpit door designed to be an impenetrable barrier against hijackers became a tool for internal catastrophe. These disasters prompted immediate regulatory changes. In the wake of the Germanwings crash, the European Aviation Safety Agency (EASA) and global regulators introduced the "two-person cockpit rule," mandating that at least two authorized personnel always be present in the cockpit. Additionally, airlines began implementing more robust psychological evaluations for pilots and reviewed internal reporting mechanisms for mental health concerns. However, the aviation industry continues to struggle with balancing mental health privacy and public safety, especially given the stigma surrounding psychological disorders in high-stakes professions.

While LAM 470 and Germanwings 9525 differ in detail, the former lacking prior mental health indicators and the latter involving known but undisclosed psychological distress, the core failure in both cases was procedural rigidity and the inability to override cockpit lockouts in emergencies. These crashes serve as sobering reminders that threats to aviation safety are not always external; sometimes, they come from within. They underscore the importance of continuous vigilance, not only in engineering and systems design but also in human factors, mental health monitoring, and procedural redundancy—the pillars of trust upon which modern aviation rests.

Automation Mismanagement and Pilot Disengagement: A Critical Examination of Air France 447, Asiana 214, and Eastern Air Lines 401

Modern commercial aircraft are equipped with sophisticated autopilot and flight management systems designed to enhance safety and reduce pilot workload. However, these systems can create dangerous vulnerabilities when pilots over-rely on automation, fail to monitor critical flight data, or misinterpret system behavior, especially during abnormal situations. Three prominent accidents—Air France Flight 447 (2009), Asiana Airlines Flight 214 (2013), and Eastern Air Lines Flight 401 (1972)—reveal the tragic consequences of automation mismanagement, pilot disengagement, and degraded situational awareness.

Air France Flight 447 (2009)

On June 1, 2009, Air France Flight 447, an Airbus A330 en route from Rio de Janeiro to Paris, crashed into the Atlantic Ocean, killing all 228 passengers and crew. The accident began with the icing of the aircraft's pitot tubes, which caused a temporary loss of reliable airspeed data. In response, the autopilot disengaged, transferring manual control to the flight crew in accordance with Airbus's design philosophy. The crew, however, failed to recognize the onset of an aerodynamic stall, despite multiple stall warnings and nose-up flight attitudes.

The BEA investigation found that the pilots misinterpreted the situation due to a lack of manual flying proficiency at high altitudes and a poor understanding of flight envelope protections. The pilot continued pulling back on the control stick, keeping the nose elevated even as the aircraft lost lift and stalled. Spatial disorientation, exacerbated by darkness and turbulence, contributed to the failure to recover (BEA, 2012). The aircraft entered a high-altitude stall from which it never recovered, descending into the ocean in a near-vertical attitude.

This accident revealed deep systemic issues with automation dependency, training deficiencies in manual flight skills, and communication breakdowns between cockpit crew members. The BEA recommended revisions in stall recognition training and emphasized the need for an improved understanding of autopilot limitations.

Asiana Airlines Flight 214 (2013)

On July 6, 2013, Asiana Airlines Flight 214, a Boeing 777 arriving from Seoul, crashed during landing at San Francisco International Airport. Three passengers died, and over 180 were injured. The aircraft approached the runway too low and too slow, ultimately striking the seawall just short of the threshold.

The NTSB investigation determined that the flight crew became confused about the behavior of the autopilot and autothrottle systems, believing that the autothrottle would maintain the target approach speed. The autothrottle had disengaged because of the manual flight mode selected during the visual approach, and the pilots failed to recognize the speed decay. Although visual conditions were good, the captain was relatively inexperienced with manual landings in the Boeing 777, and there was an evident lack of effective crew monitoring and cross-checking (NTSB, 2014).

This accident revealed the dangers of automation complacency, mode confusion, and diminished manual flying skills—issues becoming increasingly common in highly automated cockpits. It underscored the need for more rigorous manual flight training and clearer interface designs between pilots and flight control systems.

Eastern Air Lines Flight 401 (1972)

On December 29, 1972, Eastern Air Lines Flight 401, a Lockheed L-1011 TriStar, crashed into the Florida Everglades, killing 101 of the 176 people onboard. The aircraft was on final approach to Miami International Airport when the crew became distracted by a faulty nose landing gear indicator light. During this time, the autopilot was unintentionally disengaged, and the aircraft began a gradual descent that went unnoticed by the crew.

The NTSB investigation found that the entire cockpit crew was focused on troubleshooting the landing gear issue and failed to monitor flight instruments or altitude. The aircraft descended unnoticed for more than two minutes before crashing (NTSB, 1973). This tragedy was one of the earliest and most dramatic demonstrations of human error, task fixation, and the failure to manage cockpit resources effectively.

Flight 401 was a turning point in aviation history and contributed directly to the development of the Crew Resource Management (CRM) training program that emphasizes effective communication, decision-making, and workload management among flight crew members.

Corrective Action

The accidents involving Air France 447, Asiana 214, and Eastern Air Lines 401 highlight the critical dangers posed by automation mismanagement and pilot disengagement. Whether triggered by unreliable sensor inputs, mode confusion, or task fixation, each crash underscores how sophisticated technology can fail catastrophically when pilots are unprepared to assume manual control, misunderstand system behavior, or lose situational awareness.

Collectively, these tragedies have led to important industry reforms: more emphasis on manual flying skills, redesign of user interfaces to reduce mode confusion, mandatory CRM training, and better feedback systems for automation status. These events serve as sobering reminders that while automation enhances safety, it also demands a higher standard of human vigilance, competence, and teamwork.

Human Factors, CRM, and Situational Awareness Breakdown

While technology and engineering are vital pillars of aviation safety, many major accidents result not only from system failures but also from lapses in human performance, particularly failures in crew resource management (CRM), situational awareness, and communication. The following three accidents—United Airlines Flight 173 (1978), Korean Air Cargo Flight 8509 (1999), and China Airlines Flight 140 (1994)—are emblematic of the consequences that arise when cockpit crews fail to manage priorities, challenge authority, or understand flight system behavior during critical moments.

United Airlines Flight 173 (1978)

On December 28, 1978, United Airlines Flight 173, a DC-8 approaching Portland, Oregon, experienced a landing gear malfunction indication. As the crew attempted to troubleshoot the gear status, they became preoccupied, circling the city for over an hour. In doing so, they neglected to monitor fuel levels, and the aircraft eventually ran out of fuel. Despite the flight engineer's multiple warnings about dwindling fuel reserves, the captain continued focusing on the gear issue, and the aircraft crashed during emergency landing procedures, killing 10 of the 189 people on board (NTSB, 1979).

The accident exposed a significant flaw in cockpit dynamics: lack of assertive communication, task fixation, and ineffective distribution of workload. The captain, in a position of authority, failed to delegate tasks or heed the concerns of other crew members. This crash was a turning point in aviation history, prompting the development of Crew Resource Management (CRM), a set of training protocols designed to improve communication, teamwork, and decision-making in multi-crew environments.

Korean Air Cargo Flight 8509 (1999)

On December 22, 1999, Korean Air Cargo Flight 8509, a Boeing 747-200F, departed London Stansted Airport and crashed just moments after takeoff, killing all four crew members. The crash was triggered by the failure of the captain's inertial reference system (IRS), which resulted in erroneous attitude information displayed on his instruments. The first officer had accurate information on his side, but he failed to challenge the captain, who maintained control despite significant bank angle discrepancies (AAIB, 2001).

This accident revealed a deeply embedded hierarchical culture within the cockpit, where junior crew members were discouraged implicitly or explicitly from questioning senior officers. Investigators found that the captain ignored standard procedures, and the crew did not perform necessary cross-checks. The crash underscored not only a CRM breakdown but also a cultural failure to promote assertiveness and communication in the cockpit. Korean Air subsequently overhauled its safety and CRM programs, even partnering with external safety consultants to address systemic training deficiencies.

China Airlines Flight 140 (1994)

On April 26, 1994, China Airlines Flight 140, an Airbus A300-600R, crashed while attempting to land at Nagoya Airport in Japan. During the final approach, the crew initiated a go-around procedure but accidentally engaged in the takeoff/go-around (TOGA) mode while simultaneously trying to fly manually. The aircraft's autopilot remained engaged, and due to conflicting inputs between the pilot and automation, the aircraft pitched up, stalled, and crashed, killing 264 of the 271 people on board (JAAIC, 1996).

The investigation revealed confusion over the autopilot logic and a lack of understanding of how the flight control system prioritized inputs. The flight crew attempted to counteract the pitch-up with manual controls, unaware that the autopilot was resisting their efforts. A lack of assertive communication and failure to conduct a proper callout of the system mode further degraded their situational awareness.

This tragedy highlighted both CRM deficiencies and the importance of understanding automated system's behavior during abnormal flight situations. Airbus and China Airlines both responded with updated training materials and expanded education on autopilot modes, including scenarios involving mode confusion and go-around automation.

Corrective Actions

The tragedies of United 173, Korean Air Cargo 8509, and China Airlines 140 collectively illustrate how human factors, not mechanical failures, can lead to catastrophic outcomes. Common threads across these incidents include poor communication, inadequate situational awareness, and difference to authority or automation during critical flight phases. These accidents also reflect the complexity of modern aviation, where safety hinges not only on equipment but also on the cognitive, social, and cultural dynamics of the cockpit crew.

In response to these failures, airlines and regulators around the world have invested heavily in CRM training, automation literacy, and assertiveness culture. The enduring lesson is clear: even the most advanced flight systems are only as safe as the humans who operate them and the teams that support them.

These cases, like Aeroflot 593, underscore that technology alone does not guarantee safety; human factors such as vigilance, cockpit discipline, communication, and automation fluency are crucial. The aviation community has since invested heavily in CRM training, stricter access control, and better pilot-automation interface education.

References

- BEA, Accident of the Airbus A310-304 F-OGQS operated by Aeroflot near Mezhdurechensk on March 23, 1994 [translated summary]. Bureau d'Enquêtes et d'Analyses pour la sécurité de l'aviation civile, 1994.
- Interstate Aviation Committee (MAK), *Final Accident Report: Aeroflot Flight 593*, 1995, Moscow.
- International Civil Aviation Organization (ICAO), *Annex 6 to the Convention on International Civil Aviation: Operation of Aircraft*, ICAO, Montreal, 2001.
- Federal Aviation Administration (FAA), *Advisory Circular 120-51E: Crew Resource Management Training*, U.S. Department of Transportation, Washington, DC, 2004.
- BEA (Bureau d'Enquêtes et d'Analyses), Final Report on the Accident to the Airbus A320-211, D-AIPX, Operated by Germanwings on 24 March 2015, 2016.
- Namibia Directorate of Aircraft Accident Investigations (DAAI), Final Report on the Accident Involving LAM Flight TM470, 2016.
- EASA (European Union Aviation Safety Agency), *Task Force on Measures Following the Germanwings Accident: Final Report*, 2016.
- Aviation Safety Network (ASN), Accident Description: TM470, 2013.
- CNN, Germanwings Crash: Copilot 'Wanted to Destroy Plane', 2015.
- BEA, Final Report on the Accident on 1st June 2009 to the Airbus A330-203 Flight AF 447, 2012, Bureau d'Enquêtes et d'Analyses.
- NTSB, Aircraft Accident Report: Descent Below Visual Glidepath and Impact With Seawall, Asiana Airlines Flight 214, San Francisco, July 6, 2013, 2014, NTSB/AAR-14/01.
- NTSB, Aircraft Accident Report: Eastern Air Lines Flight 401, Lockheed L-1011, December 29, 1972, 1973.
- NTSB, Aircraft Accident Report: United Airlines Flight 173, DC-8-61, Portland, Oregon, December 28, 1978, 1979.
- UK Air Accidents Investigation Branch (AAIB), Report on the Accident to Korean Air Cargo Flight 8509, Boeing 747-2B5F, G-BMGS, near London Stansted, December 22, 1999, 2001.
- Japan Aircraft Accident Investigation Commission (JAAIC), Final Report on the Accident to China Airlines Flight 140, Airbus A300B4-622R, Nagoya Airport, April 26, 1994, 1996.

19

Calibration, Culture, and the Courage to Question

Abstract
The illusion of precision in measurement systems conceals deeper cultural failures. True traceability demands a mindset shift from compliance to accountability. Systems must empower questioning, not just adherence to standards. Failures often stem from silence, where potential issues go unreported due to hierarchical barriers. Cultivating a culture of inquiry, rewarding dissent, and embedding feedback loops are essential. Accuracy is an ongoing pursuit, requiring humility and vigilance. To advance reliability, prioritize open communication, continuous learning, and a commitment to challenge assumptions. Let courage in questioning be the standard for calibration.

***Keywords*:** Food safety, quality control, nutritional accuracy, packaging standards, shelf life, regulatory compliance, consumer trust

The Illusion We Built

The preceding chapters have illuminated a fundamental truth often obscured in the fog of process charts, equipment certifications, and audit checklists: the illusion of precision is not precision. Behind every digital readout, beneath every metrological standard, there lies a human story of oversight, adaptation, silence, or courage. When traceability fails, it is not merely because a measurement lost its tether to a standard, but because a conversation did not happen, a challenge was not voiced, or a doubt was buried under the weight of routine. We build systems to make complexity manageable, but in doing so, we sometimes mask the very risks we hope to control. And when the illusion of accuracy becomes more comforting than the pursuit of truth, we set the stage for failure.

Michael D. Holloway. Accuracy and Calibration Failures in Science and Engineering, (259–262)

From Compliance to Culture

Compliance-based systems are the bedrock of many industries. Regulations demand checklists, certifications, and documented calibrations. Yet time and again, history shows us that compliance alone is not enough. The most catastrophic failures, from Three Mile Island to the Deepwater Horizon and from Challenger to Chernobyl, did not occur because people failed to fill out paperwork. They happened because of cultural failures: unquestioned assumptions, normalized deviance, and the subtle erosion of vigilance over time.

True traceability is not just a paper trail. It is a mindset. It means creating systems where calibration is not just scheduled, it is understood. Where measurements are not just recorded but questioned. Where people are not just trained to follow, they are empowered to ask. Traceability in culture means ensuring every operator, technician, scientist, or engineer sees themselves as a guardian of accuracy, not a passive participant in its decay.

This cultural mindset requires a shift from external enforcement to internal accountability. It requires leaders who model curiosity, humility, and an intolerance for comfortable assumptions. It requires peer networks that value reflection as much as results. And above all, it requires recognizing that measurement systems are fallible not because they are bad, but because they are human.

The Courage to Ask, Challenge, and Speak Up

Every failure begins as silence. Someone noticed something but did not report it. Someone who had a hunch but lacked the authority to raise it. Someone who knew the risk but assumed someone else would speak up. This culture of quiet compliance is not merely dangerous; it is deadly.

Cultivating the courage to question is one of the most underappreciated skills in scientific and engineering domains. It takes bravery to ask, "How do we know this is still true?" It takes integrity to say, "This number does not look right, even if it is within spec." It takes resolve to challenge consensus, especially in environments where hierarchy or reputation rules. And yet, this is where safety lives.

Organizations must do more than create open-door policies or suggestion boxes. They must embed the expectation to question into their very operating DNA. From design reviews to calibration verifications, from data audits to root cause investigations, systems must reward dissent, not just agreement. Peer review must be more than a formality. Verification must be more than a form. And feedback must be a mechanism for growth, not punishment.

The courage to challenge cannot live in individuals alone. It must be institutionalized, trained, modeled, and celebrated. It must be seen not as insubordination, but as the highest form of professionalism.

Pull It Together

In the end, the failures explored in this book, whether technological, institutional, or personal, reveal a deeper truth: traceability is not only a metrological requirement but also a cultural one. We lose our way not just when standards shift or instruments drift, but when people stop asking questions, when institutions suppress doubt, and when systems are designed to reward certainty over scrutiny. The cost of error is rarely just data; it is safety, credibility, trust, and sometimes lives. To prevent these failures, we must embed a culture of inquiry, not compliance; one that treats calibration not as a checkbox but as an evolving responsibility. This means designing feedback loops into systems, elevating the role of human factors, and cultivating environments where dissent is not only allowed but essential. Only when we recognize that accuracy is a pursuit, not a possession, can we begin to build organizations worthy of trust. If the illusion of accuracy brought us here, then only humility and vigilance will carry us forward.

A Call to Action

What Will You Do Differently Tomorrow?

- **Audit one assumption** you've been making without verification.
- **Review one measurement system** for hidden drift, outdated calibration, or uncontrolled changes.
- **Encourage one teammate** to ask a difficult question.
- **Report one near miss or inconsistency** that has been left undocumented.
- **Schedule one conversation** with your team or supervisor about "how we know what we know."
- **Teach one principle** from this book to someone newer in your organization.

- **Review your role** in your company's calibration culture and identify one area you can improve.
- **Speak up once** when it is uncomfortable but necessary.

A Final Note

This book was not written to indict science or engineering, but to illuminate the fragile brilliance of these disciplines. They succeed not because they are perfect, but because they are self-correcting. That correction, however, only happens when people are willing to trace the error back to its root and then have the courage to act on what they find.

Let that courage be your calibration standard.

Appendix

Examples of SOPs, Dashboards, or Audit Forms

Title: *SOP-QC-001 – Calibration and Use of Analytical Balances*

Purpose:
To establish a standardized method for calibrating and verifying analytical balances used in laboratory and production processes.

Scope:
Applies to all balances with ≥0.1 mg readability used for material weighing.

Responsibilities:

- Lab technician: Perform daily verification and document results.
- QA specialist: Review calibration logs monthly.
- Metrology team: Perform external calibration annually.

Procedure Summary:

1. **Daily Verification**
   - Use certified Class E2 weights traceable to NIST.
   - Perform two-point verification (e.g., 1 g and 100 g).
   - Document on balance verification log (Form F-QC-002).
2. **Monthly Review**
   - QA to review deviation trends and report anomalies.
   - Out-of-tolerance readings require quarantine and RCA.
3. **Annual Calibration**
   - External ISO 17025–certified service provider.
   - Receive and file calibration certificate with uncertainty budget.

Michael D. Holloway. Accuracy and Calibration Failures in Science and Engineering, (263–266)

4. **Out-of-Tolerance Protocol**
   - Immediate notification to QA and halt of affected processes.
   - Retrospective impact assessment on product batches.

Example Dashboard: Calibration Status Interface

A **real-time calibration dashboard** visualized through a Quality Management System (QMS) or asset management software enables immediate visibility into instrument readiness.

Dashboard Features:

Instrument ID	Type	Last calibrated	Next due	Status	Assigned to	Notes
BAL-2034	Analytical balance	2024-11-15	2025-11-15	GREEN In Spec	Tech 1	Annual calibration complete
PH-1077	pH meter	2024-12-02	2025-06-02	YELLOW Due Soon	Tech 2	Buffer stock low
GC-4500	GC-FID	2024-05-01	2025-05-01	GREEN In Spec	Lab 3 Lead	Drift within expected range
PRESS-118	Pressure transducer	2023-12-10	2024-12-10	RED Overdue	Metrology	Quarantine initiated

Color Coding:

- Green = In Spec
- Yellow = Due Soon (within 30 days)
- Red = Overdue – must not be used

KPI Metrics Displayed:

- % of instruments calibrated on time
- Number of OOT (out-of-tolerance) events this quarter
- Instruments in quarantine
- Technicians flagged for recurring late verification

Example Internal Audit Form: Calibration Integrity Checklist

Title: *QA Internal Audit – Calibration Program Review (Form A-QA-018)*
Frequency: Semiannual
Auditor: [Name]
Date: [Insert]

Audit area	Question	Yes/no	Findings/ comments
Inventory accuracy	Is the full instrument list current and verified?		
Calibration traceability	Are all records traceable to national/international standards?		
Status labeling	Are all instruments clearly labeled with calibration status?		
Verification logs	Are daily/weekly verifications performed and documented?		
Out-of-Tolerance Handling	Are OOT events properly investigated and escalated?		
Certificate review	Are all external calibrations reviewed for uncertainty budgets?		
Training and Awareness	Do users understand calibration impact and SOP expectations?		
SOP compliance	Are procedures followed, and deviation reports filed when needed?		
Digital system integration	Is calibration status integrated into dashboards or LIMS/ QMS?		
Continuous improvement	Are lessons from calibration failures incorporated into system updates?		

Audit Summary:

- Findings:
- Corrective actions required:
- Due dates & responsible parties:

Implementation Notes

- These tools are **modular and scalable**, designed for manufacturing, analytical labs, or R&D settings.
- All documents should align with your organization's **QMS, document control protocols, and regulatory context** (e.g., FDA, ISO, EPA, and GxP).
- Audit findings should feed into **management review meetings**, and dashboards should be visible to operators and QA in real time.
- SOPs and dashboards should be **linked**, so overdue status triggers SOP-defined actions.

References

- ISO/IEC 17025:2017, General Requirements for the Competence of Testing and Calibration Laboratories.
- USP <41> – Balances and <1251>, Weighing on an Analytical Balance.
- FDA 21 CFR Part 11, Electronic Records; Electronic Signatures.
- NIST Handbook 44, Specifications, Tolerances, and Technical Requirements for Weighing Devices.
- CLSI QMS06, Equipment Calibration and Maintenance.
- ILAC G8:09/2019, Guidance on Decision Rules and Statements of Conformity.

Measurement Integrity Readiness Index (MIRI)

A Calibration Culture Maturity Model for Technical Organizations

The **Measurement Integrity Readiness Index (MIRI)** is a five-level model that evaluates an organization's calibration and measurement practices across technical, cultural, and leadership dimensions. Each level reflects an organization's ability to **detect, respond to, and learn from data uncertainty, calibration drift, and metrological failure**.

MIRI Levels Overview

Level	Maturity stage	Description
1	Reactive	Calibration is compliance-driven and reactive; drift is often unnoticed.
2	Defined	SOPs exist, but integration, training, and culture are inconsistent.
3	Managed	Calibrations are monitored systematically; some visibility and feedback loops.
4	Proactive	Calibration linked to risk, quality, and operational metrics; continuous learning.
5	Embedded & ethical	Integrity culture is system-wide; calibration informs strategy and leadership.

Michael D. Holloway. Accuracy and Calibration Failures in Science and Engineering, (267–272)

Assessment Dimensions and Indicators

Dimension	Key indicators	Level 1	Level 3	Level 5
Leadership engagement	Executive sponsorship, messaging, and budget for calibration integrity.	Absent	Involved	Strategic
SOP & traceability	Documented calibration procedures with traceability to national/ international standards.	Informal	Standardized	Audited and optimized
Data visibility	Real-time dashboards and calibration status linked to operations.	Manual logs	Partial	Integrated
Training & literacy	Calibration training for technicians, QA, and cross-functional teams.	None/basic	Periodic	Continuous
Risk & RCA integration	Use of calibration data in root cause analysis and preventive actions.	Rare	Case-based	Systematic
Cultural safety	Staff comfort in reporting out-of-tolerance conditions or failures.	Fearful	Cautious	Safe and encouraged
Digital systems	Use of LIMS, QMS, or CMMS to track, flag, and analyze calibration status.	Spreadsheet	Mixed	Real-time and automated
Response to Drift	Behavior when drift is detected: correction, escalation, impact assessment.	Ad hoc	SOP-guided	Preventive and transparent

(*Continued*)

(*Continued*)

Dimension	Key indicators	Level 1	Level 3	Level 5
Ethical accountability	Calibration data used to make ethical, patient- or customer-centric decisions even at a business cost.	Ignored	Considered	Actively modeled by leadership

How to Use the MIRI

1. **Self-Assessment:** Score your organization across each dimension (1–5).
2. **Profile Your Calibration Culture:** Identify where you fall—reactive, defined, managed, proactive, or embedded.
3. **Targeted Action Planning:** For dimensions scoring ≤3, develop short-term and long-term corrective actions (e.g., new SOPs, leadership briefings, and software upgrades).
4. **Track Progress Over Time:** Reassess annually; integrate with internal audits or quality management reviews.

Example Scoring Template (Condensed)

Dimension	Score (1–5)	Action priority	Notes/gaps identified
Leadership engagement	3	Medium	Budget allocated, but not championed
Data visibility	2	High	No integration with QMS or SCADA
Risk & RCA integration	4	Low	Used regularly after major incidents

Outcome Use Cases

- **For Regulators:** Evaluate organizational readiness beyond certificate compliance.
- **For Executives:** Benchmark calibration maturity against peers and industry best practices.

- **For QA/QC Teams:** Prioritize investments in training, software, or audit enhancements.
- **For Consultants:** Structure improvement roadmaps and performance KPIs.

References

- ISO/IEC 17025:2017, General Requirements for the Competence of Testing and Calibration Laboratories.
- ISO 9001:2015, Quality Management Systems – Requirements.
- CLSI QMS06, Equipment Calibration and Maintenance.
- FDA 21 CFR Part 11, Electronic Records; Electronic Signatures.
- Schein, E.H., *Organizational Culture and Leadership*, 2010.
- Edmondson, A., *The Fearless Organization*, 2019.
- ILAC G8:09/2019, Statements of Compliance and Decision Rules.
- ASQ, *Measurement System Analysis (MSA) Manual*, 4th, 2020.

Industry Standards and Definitions

Core Metrology and Calibration Standards

- SI Units (Système International d'Unités)
  The global system of measurement units (e.g., meter, second, and kilogram) is recognized under the International System of Units. All traceable measurements should be linked to SI units to ensure consistency across space and time.
- ISO/IEC 17025
  General Requirements for the Competence of Testing and Calibration Laboratories Specifies the competence of laboratories to carry out calibration and testing, including data integrity, documentation, traceability, and uncertainty reporting.
- NIST Traceability
  A measurement is traceable to NIST when it can be linked to national standards *via* an unbroken chain of documented comparisons, each with stated uncertainty. Ensures international comparability and reliability of results.
- SRM (Standard Reference Material)
  Certified materials issued by NIST and similar bodies are used to calibrate instruments and validate methods.

Example: SRM 2806a for particle counting calibration in lubricants.

Data Quality and Analytical Standards

- GxP (Good Practice) Standards
  Includes GLP (Good Laboratory Practice), GMP (Good Manufacturing Practice), and GCP (Good Clinical Practice). These frameworks enforce quality control, audit trails, and data integrity, especially in life sciences and pharmaceuticals.
- ISO 9001
  Quality Management Systems – Requirements Establishes the criteria for a quality management system (QMS), including documentation, corrective actions, and continual improvement based on measurable data.
- ALCOA+ Principles
  Data integrity model requiring that records be:
  - Attributable
  - Legible
  - Contemporaneous
  - Original
  - Accurate

    Plus: Complete, consistent, enduring, and available
- Uncertainty of measurement
  A parameter associated with the result of a measurement, characterizing the dispersion of values that could reasonably be attributed to the measurand. Critical for comparing measurements and making decisions under uncertainty.

Risk, Reliability, and Failure Analysis Standards

- FMEA (Failure Modes and Effects Analysis)
  A structured method for evaluating and prioritizing risks in a system based on the severity, occurrence, and detectability of potential failures.
- HAZOP (Hazard and Operability Study)
  A team-based risk assessment approach is used to identify hazards and operability issues in process systems by examining deviations from design intent.

- Root Cause Analysis (RCA)
  A systematic process for identifying the underlying causes of an event or failure, not just the symptoms, to prevent recurrence.
- ISO 31000
  Risk Management – Guidelines
  Offers principles and a framework for managing risk in any organization, emphasizing integration into all decision-making levels.

Environmental and Laboratory-Specific Standards

- EPA 40 CFR Part 136
  U.S. EPA guidelines for test procedures used in analyzing pollutants in water under the Clean Water Act.
- ASTM D7647/ISO 11171
  Standards for calibrating automatic particle counters used in fluid power systems. Crucial for tracking oil cleanliness and lubricant degradation.
- ISO 4406
  Hydraulic fluid power – Fluids – Method for coding the level of contamination by solid particles using automatic particle counters.

Human Factors, Interface, and Communication Standards

- ANSI/HFES 100
  Human Factors Engineering of Computer Workstations Standards for designing ergonomic, cognitively aligned interfaces that reduce operator error and fatigue.
- IEC 61508
  Functional Safety of Electrical/Electronic/Programmable Systems Covers risk assessment and reliability requirements for safety-related control systems, including human interface considerations.
- NTSB Safety Culture Guidelines
  Although not a formal standard, these guidelines emphasize blameless reporting, cross-hierarchy communication, and learning from near misses, which are critical for a culture of safety and continuous improvement.

Terminology

Term	Definition
Alarm fatigue	A state in which operators become desensitized to frequent or non-critical alerts, increasing the risk of missing critical warnings.
Analytical blind spot	A hidden area of vulnerability in a system or model where data is not collected, interpreted, or validated leading to incorrect or incomplete conclusions.
Assumption drift	The gradual divergence between the assumptions a system or model was designed on and the evolving real-world conditions it is applied to.
Audit trail	A verifiable record of the steps taken in data collection, analysis, model development, and calibration, enabling traceability and accountability.
Bias (cognitive or model-based)	Systematic deviation in judgment or data interpretation, often rooted in flawed heuristics or model training on incomplete data.
Blameless postmortem	A retrospective investigation of failures or near misses that focuses on systemic improvement rather than individual fault.
Calibration drift	A slow, often undetected change in an instrument's accuracy or performance over time, which may lead to incorrect measurements if uncorrected.
Calibration lifecycle	The complete process of an instrument's calibration, from initial configuration through ongoing recalibration, performance checks, and retirement.
Challenge loop	A feedback mechanism built into decision-making and system design that encourages dissent, red-teaming, and model stress testing to uncover hidden flaws.

(*Continued*)

Michael D. Holloway. Accuracy and Calibration Failures in Science and Engineering, (273–276)

(*Continued*)

Term	Definition
Confidence interval	A statistical range that expresses the uncertainty or variability in a measurement or forecast, typically accompanied by a confidence level (e.g., 95%).
Cross-functional integration	The active collaboration between different departments (e.g., technical, operational, executive) to ensure decisions are balanced, transparent, and actionable.
Cross-validation (models)	A method of testing predictive models by dividing data into training and validation sets to assess generalizability and robustness.
Data drift	A change in the statistical properties of input data over time, leading to degradation in model accuracy or decision quality.
Data integrity	The completeness, accuracy, consistency, and reliability of data throughout its lifecycle, from acquisition to analysis to storage.
Data provenance	Documentation of the origin, transformations, and custody of data essential for validation and trust in complex analytical environments.
Devil's advocate review	A formal role or process designed to critically challenge group consensus, surface dissenting viewpoints, and reduce groupthink.
Discrepancy flagging	A protocol for identifying and surfacing mismatches between redundant systems, model outputs, or expected vs. actual values.
Failure mode	The specific way in which a system, component, or process can fail, including both physical breakdowns and decision failures.
False confidence	The unjustified certainty in a system, model, or measurement, often caused by lack of uncertainty reporting or overfitting to historical success.
Historical data normalization	The process of adjusting older data sets to be compatible with new measurement standards, calibration curves, or system architectures.

(*Continued*)

(*Continued*)

Term	Definition
Human factors	The design and operational considerations that account for human strengths, limitations, decision-making, and error tendencies in complex systems.
Interface ambiguity	A user interface flaw where system displays fail to clearly convey the true state of the system, leading to misinterpretation or delayed action.
Latent risk	A hidden vulnerability in a system that does not immediately manifest but can contribute to failure under specific conditions.
Metrological traceability	The property of a measurement to be linked to a reference standard (typically SI units) through an unbroken chain of comparisons, each with a documented uncertainty.
Model overfitting	A modeling error in which a system is too closely tuned to historical data, reducing its ability to generalize to new or unexpected situations.
Normalization of deviance	The gradual acceptance of substandard performance or policy violations as "normal" due to the absence of immediate consequences.
Operational drift	Incremental departure from standard operating procedures or design limits, often unnoticed until a major failure occurs.
Red team	An independent group tasked with stress-testing a system, protocol, or plan by adopting an adversarial or contrarian perspective.
Resilience engineering	A discipline focused on designing systems that can adapt, recover, and learn from disruptions, failures, or unforeseen challenges.
Retrospective analysis	The post-event review of decisions and system behavior, often using a structured framework to uncover contributing factors and hidden risks.

(*Continued*)

(*Continued*)

Term	Definition
Risk register	A centralized, living document used to log, assess, prioritize, and track identified risks, mitigation actions, and outcomes.
Root cause analysis (RCA)	A structured method to identify the fundamental cause of a problem or failure, rather than its symptoms.
Self-diagnostic system	A system that continuously monitors its own performance, alerts for calibration drift, or flags invalid measurement states.
Sensitivity analysis	A technique used to determine how different values of input variables affect a model's output, often used to identify high-risk assumptions.
Scenario testing	The use of constructed, often extreme "what-if" cases to test the behavior, performance, and robustness of models, systems, or response plans.
Sensor fusion	The use of multiple sensor inputs often from diverse modalities to improve accuracy, reduce uncertainty, and enable cross-verification.
Signal-to-noise ratio	A measure of how clearly a true signal (measurement or data pattern) can be distinguished from background noise or error.
Systemic risk	The risk of failure not from an isolated flaw but from the interaction of multiple components or policies within a system or network.
Uncertainty budget	A detailed accounting of all known sources of uncertainty in a measurement, model, or system output.
Version control (models/data)	The practice of tracking and managing changes to datasets, code, calibration curves, or algorithms to ensure transparency, reproducibility, and accountability.
Worst-reasonable-case	A scenario that is unlikely but still plausible, used to test system design, emergency protocols, and leadership decision-making capacity.

Templates

Postmortem Review

Use this structured format following a failure, anomaly, or high-risk deviation. It encourages blameless, constructive analysis and helps surface systemic, cultural, and technical contributing factors.

1. Summary of the Event
   - Title/description of incident:
   - Date/time of occurrence:
   - System/process involved:
   - Reported by:
   - Severity level:
     ☐ Near miss ☐ Minor impact ☐ Major impact
     ☐ Catastrophic
2. What Happened? (Factual Timeline)
   - Sequence of events:
     (List chronological steps leading up to, during, and after the event.)
   - Key triggers or indicators:
     (What warnings or data points occurred? Were they seen/interpreted?)
3. Technical Root Cause(s)
   - ☐ Sensor error or drift
   - ☐ Model misclassification
   - ☐ Interface ambiguity
   - ☐ Calibration or standard mismatch
   - ☐ Algorithmic flaw
   - ☐ External/environmental factor
   - ☐ Other: ______________________________
4. Human/Systemic Contributors
   - ☐ Communication breakdown
   - ☐ Hierarchy suppression
   - ☐ Training deficiency

Michael D. Holloway. Accuracy and Calibration Failures in Science and Engineering, (277–282)

- ☐ Poor assumptions
- ☐ Overconfidence/complacency
- ☐ Data gap or loss of traceability
- ☐ Missing or unheeded change detection
- ☐ Unvalidated model edge cases

5. Impact Assessment
   - Technical or operational consequences:
   - Customer or stakeholder impact:
   - Financial cost:
   - Compliance or safety exposure:
6. What Could Have Prevented It?
   - (What signals were missed? What systems failed silently? What safeguards were lacking?)
7. Recommended Corrective Actions

Action	Owner	Due date	Status

8. Follow-Up and Training Integration
   - Has the lesson been added to the Lessons Learned Register? ☐ Yes ☐ No
   - Will it be used in future training or simulation exercises? ☐ Yes ☐ No
   - Required SOP/policy/interface revisions:_______________

Audit Checklist for Measurement and Calibration Integrity

Use this checklist for quarterly or annual audits of systems involving sensors, models, calibrations, or decision automation.

A. Calibration and Traceability
   - ☐ Are all critical instruments within calibration dates?
   - ☐ Is calibration traceable to SI units (*via* ISO/IEC 17025 labs or national standards)?
   - ☐ Are calibration certificates stored and reviewed during data disputes?

- ☐ Is there automated logging of calibration versions or adjustments?
- ☐ Are uncertainty budgets available for key measurement systems?

B. Data Quality and Integrity

- ☐ Is raw data clearly linked to source and method (data provenance)?
- ☐ Are assumptions, limitations, and confidence intervals disclosed with outputs?
- ☐ Is a data drift detection system in place for AI/ML or time-series systems?
- ☐ Are historical data sets flagged for pre/post calibration shifts?

C. Human Factors and Interface Design

- ☐ Do interfaces clearly display system state (normal, error, degraded)?
- ☐ Can users distinguish estimated vs. measured values?
- ☐ Are critical alerts visually prioritized?
- ☐ Are system users trained on skepticism and failure recognition, not just operation?

D. Change Management

- ☐ Are there versioned records for:
  - ☐ Calibration standard changes
  - ☐ Algorithm or model updates
  - ☐ Instrument firmware/software upgrades
- ☐ Is a change impact assessment conducted when any core measurement method is altered?
- ☐ Are stakeholders formally notified of calibration/model updates?

E. Culture and Safeguards

- ☐ Is a Lessons Learned Register maintained and reviewed regularly?
- ☐ Have simulations or drills been run for recent worst-case scenarios?
- ☐ Are cross-disciplinary review boards active and documented?
- ☐ Are red teams or devil's advocates used before major system rollouts?
- ☐ Are there anonymous channels for dissent or whistleblowing?

F. Score Summary

Category	Total items	Items checked	Compliance %
Calibration & traceability	5	___	___%
Data quality & integrity	4	___	___%
Human factors	4	___	___%
Change management	4	___	___%
Culture & safeguards	5	___	___%
Total	22	___	___%

Calibration Drift Audit Checklist

Guide to Identifying, Assessing, and Mitigating Drift in Measurement Systems

I. **Documentation and Traceability**
- ☐ Are all instruments on the calibration schedule documented with unique identifiers?
- ☐ Is there a clear record of the last calibration date, calibration standard, and service provider?
- ☐ Are calibration certificates traceable to **national or international standards** (e.g., NIST, ISO 17025)?
- ☐ Are records digitally archived and protected against alteration or loss?
- ☐ Is there a procedure for handling expired or out-of-tolerance calibration certificates?

II. **Calibration History and Trend Analysis**
- ☐ Are calibration results trended over time for each instrument?
- ☐ Are deviations from nominal or previous values flagged for review?
- ☐ Has the frequency of required recalibration increased for any instruments?

- ☐ Are instruments that frequently require adjustment flagged for potential replacement or root cause analysis?
- ☐ Is drift history used to adjust calibration intervals (i.e., predictive instead of static scheduling)?

III. Environmental and Operational Factors

- ☐ Are temperature, humidity, and vibration exposure monitored in measurement environments?
- ☐ Are instruments used outside their specified environmental range flagged for increased calibration risk?
- ☐ Are mobile or field-deployed instruments subject to more frequent recalibration checks?
- ☐ Are operators trained to recognize conditions that could accelerate drift?

IV. Procedures and Verification Practices

- ☐ Are periodic intermediate checks (e.g., using reference materials or known standards) in place between formal calibrations?
- ☐ Are operators required to perform verification checks before critical measurements?
- ☐ Are out-of-tolerance findings documented with corrective actions and impact assessments?
- ☐ Is there a defined escalation process when verification fails or drift is suspected?

V. Training and Human Factors

- ☐ Are personnel trained to understand calibration drift and its impact on data quality?
- ☐ Are they empowered to report anomalies without fear of reprisal?
- ☐ Is psychological safety encouraged to ensure open reporting of suspected drift or misuse?
- ☐ Are equipment users clearly distinguished from those responsible for calibration?
- ☐ Is cross-training in place to reduce dependency on a single calibration technician?

VI. System Integration and Digital Alerts

- ☐ Are calibration statuses visible in production or lab dashboards?
- ☐ Do software systems flag instruments that are out of calibration or nearing expiration?
- ☐ Are automatic lockouts or alerts used to prevent data collection from non-compliant instruments?

- ☐ Are audit trails for changes in calibration intervals, procedures, or instruments maintained?

VII. Quality System and Audit Readiness

- ☐ Does the internal audit plan include checks for calibration drift trends and gaps?
- ☐ Are calibration failures tied into CAPA (Corrective and Preventive Action) systems?
- ☐ Is calibration effectiveness reviewed during management reviews or quality meetings?
- ☐ Have external audit findings ever cited calibration drift or lapse in control?
- ☐ Are vendors and subcontractors evaluated for their calibration drift controls?

VIII. Drift Mitigation Strategy

- ☐ Is there a formal calibration drift mitigation plan?
- ☐ Are critical instruments assessed for redundancy or backup verification?
- ☐ Is risk-based calibration interval adjustment used (e.g., FMEA, reliability modeling)?
- ☐ Are drift-prone instruments prioritized for replacement or redesign?
- ☐ Is there a plan for root cause analysis if measurement error is traced to drift?

Scoring and Risk Prioritization (Optional Tool)

For each section, rate compliance:

- **3 – Fully compliant**
- **2 – Partially compliant**
- **1 – Non-compliant or unknown**

Use the total score to prioritize corrective actions and highlight high-risk areas.

Detailed Checklists by Industry

Pharmaceutical & Biotech Industry

Focus Areas: GMP compliance, FDA/EMA audits, validated equipment
Checklist:

- ☐ Are all critical instruments included in the validation master plan (VMP)?
- ☐ Are calibration records included in batch release documentation?
- ☐ Are stability chambers, autoclaves, and incubators monitored for drift *via* continuous data loggers?
- ☐ Are high-purity HPLC/GC instruments verified using system suitability tests and certified reference standards?
- ☐ Is instrument drift evaluated during OOS (out-of-specification) investigations?
- ☐ Are calibration SOPs aligned with FDA 21 CFR Part 11 and USP <1058>?
- ☐ Is there a CAPA process in place for calibration failures affecting validated methods?

Relevant Regulations: FDA 21 CFR 211, EMA Annex 11, USP <1058>, ISO 17025
Pharmaceutical and Biotech Reference

- **USP <1058>** – *Analytical Instrument Qualification*
  U.S. Pharmacopeia guidance for instrument lifecycle, calibration, and verification.
- **FDA 21 CFR Part 11 & Part 211** – *Electronic Records; Current Good Manufacturing Practice for Finished Pharmaceuticals*
  U.S. regulatory requirements for equipment calibration, maintenance, and data integrity.

Michael D. Holloway. Accuracy and Calibration Failures in Science and Engineering, (283–290)

- **EMA Annex 11** – *Computerised Systems*
  European guidelines for electronic system validation and calibration traceability.

Aerospace & Defense

Focus Areas: Safety-critical systems, traceability to NIST, AS9100 compliance
Checklist:

- ☐ Are torque wrenches, micrometers, pressure transducers, and altimeters recalibrated per OEM or military specs?
- ☐ Are calibration certificates traceable to national standards (e.g., NIST and UKAS)?
- ☐ Are test benches and avionics simulation devices subject to periodic inter-lab cross-verification?
- ☐ Are non-conforming instruments quarantined and subject to root cause analysis?
- ☐ Is calibration data used in Failure Modes and Effects Analysis (FMEA)?
- ☐ Are instrument tolerances and drift rates included in design validation plans?

Relevant Regulations: AS9100, NADCAP, MIL-STD-45662A, ISO 10012
Aerospace and Defense Reference

- **AS9100D** – *Quality Management Systems – Requirements for Aviation, Space, and Defense Organizations*
  Requires calibration systems that ensure equipment accuracy and reliability.
- **MIL-STD-45662A** – *Calibration Systems Requirements*
  Former U.S. military standard still referenced in legacy contracts.
- **NADCAP AC7004** – *Quality System Requirements for Suppliers of Calibration Services*

Oil & Gas/Petrochemicals

Focus Areas: Process safety, custody transfer, environmental compliance

Checklist:

- ☐ Are flow meters (e.g., Coriolis and ultrasonic) and pressure transmitters calibrated per API/ASTM standards?
- ☐ Are gas chromatographs and spectrometers checked against certified gas/liquid standards?
- ☐ Is calibration drift monitored in SIS (safety instrumented systems)?
- ☐ Are mass balance discrepancies used as drift indicators in custody transfer?
- ☐ Are shutdowns or flaring incidents investigated for measurement error due to drift?
- ☐ Is calibration alignment between field devices and DCS/SCADA systems verified?

Relevant Standards: API MPMS, ASTM D6299, IEC 61511, ISO 5168
Oil & Gas/Petrochemicals Reference

- **API MPMS (Manual of Petroleum Measurement Standards)**
  Multiple chapters detailing calibration of flow meters, analyzers, and tank gauging systems.
- **ASTM D6299-13** – *Standard Practice for Applying Statistical Quality Assurance and Control Charting Techniques to Evaluate Analytical Measurement System Performance*
- **IEC 61511** – *Functional Safety – Safety Instrumented Systems for the Process Industry Sector*

Environmental Testing & Monitoring

Focus Areas: Regulatory reporting, traceability, EPA/ISO methods
Checklist:

- ☐ Are air, water, and soil samplers calibrated using EPA-approved standards?
- ☐ Are calibration records for continuous emissions monitoring systems (CEMS) retained per retention requirements?
- ☐ Are GC-MS and ICP-OES instruments recalibrated using NIST-traceable calibration standards?
- ☐ Are sampling errors or instrument drift factored into environmental impact assessments?

- ☐ Is drift evaluated during method detection limit (MDL) studies?
- ☐ Are all changes to calibration protocols documented per ISO 17025?

Relevant Standards: EPA Method 6020B, ISO 14001, NELAC, ISO/IEC 17025

Environmental Testing and Monitoring Reference

- **EPA Method 6020B** – *Inductively Coupled Plasma-Mass Spectrometry*
  Requires routine calibration verification using traceable standards.
- **NELAC Institute Standards** – *National Environmental Laboratory Accreditation Program (TNI)*
- **ISO 14001:2015** – *Environmental Management Systems – Requirements*

Food & Beverage Manufacturing

Focus Areas: HACCP, FSMA compliance, product safety
Checklist:

- ☐ Are temperature probes in pasteurizers, chillers, and ovens regularly calibrated?
- ☐ Are pH meters and moisture analyzers verified against standard buffers/materials?
- ☐ Is equipment calibration tied directly to CCPs (Critical Control Points) in HACCP plans?
- ☐ Are scales used in formulation and packaging calibrated and certified?
- ☐ Are sanitation and cleaning verification systems (e.g., ATP meters) validated and drift-checked?
- ☐ Are calibration errors documented as part of non-conformance reports?

Relevant Standards: HACCP, FSMA, ISO 22000, USDA Guidelines

Food & Beverage Industry Reference

- **HACCP (Hazard Analysis and Critical Control Points)** – Codex Alimentarius Calibration of monitoring equipment at CCPs is mandatory.
- **FSMA (Food Safety Modernization Act)** – U.S. FDA Requires documentation and calibration of process monitoring tools.
- **ISO 22000:2018** – *Food Safety Management Systems – Requirements*

Automotive & Manufacturing Industry

Focus Areas: ISO/TS 16949 compliance, dimensional measurement, traceability
Checklist:

- ☐ Are CNC machine probes, coordinate measuring machines (CMMs), and gauges calibrated at defined intervals?
- ☐ Is gauge R&R (repeatability and reproducibility) performed to track long-term instrument behavior?
- ☐ Are tool wear and fixture deformation checked as part of drift analysis?
- ☐ Is there a digital record of all gauge history and calibration events?
- ☐ Are control charts used to track measurement drift or offset?
- ☐ Are torque tools verified before assembly line use?

Relevant Standards: IATF 16949, ISO 10012, AIAG MSA
Automotive and Manufacturing Reference

- **IATF 16949:2016** – *Automotive Quality Management Standard*
  Includes requirements for equipment calibration and verification.
- **AIAG MSA (Measurement Systems Analysis Manual, 4th Edition)** – Automotive Industry Action Group

Clinical & Medical Laboratories

Focus Areas: Patient safety, CLIA/CAP accreditation, method validation
Checklist:

- ☐ Are pipettes, spectrophotometers, and analyzers regularly calibrated and verified using traceable reference materials?
- ☐ Are calibration verification logs part of each method validation report?
- ☐ Is drift in instruments like hematology analyzers or blood gas machines monitored using control charts?
- ☐ Is any failed calibration immediately evaluated for patient impact or test invalidation?
- ☐ Is calibration reviewed during routine proficiency testing and CAP inspections?
- ☐ Are software updates evaluated for their effect on calibration?

Relevant Regulations: CLIA, CAP Checklists, ISO 15189, CLSI QMS06
Clinical and Medical Laboratories Reference

- **CLIA (Clinical Laboratory Improvement Amendments)**
  U.S. federal standards for laboratory testing on human specimens.
- **CAP Accreditation Checklists** – *College of American Pathologists*
  Includes calibration verification, linearity, and instrument drift control.
- **ISO 15189:2022** – *Medical laboratories – Requirements for quality and competence*
- **CLSI QMS06-A** – *Application of a Quality Management System Model for Laboratory Services*

Semiconductor and Precision Electronics

Focus Areas: Nano-scale measurement, tight tolerances, equipment aging
Checklist:

- ☐ Are coordinate measuring machines (CMMs), profilometers, and interferometers calibrated with sub-micron accuracy?

- ☐ Is ESD (electrostatic discharge) monitoring equipment regularly verified?
- ☐ Are thermal drift effects on lithography and etch tools compensated and logged?
- ☐ Are probe station measurements traceable and monitored for long-term offset?
- ☐ Is drift evaluated using SPC tools across wafer lots?
- ☐ Are environmental controls (cleanroom temp/humidity) tied to calibration protocols?

Relevant Standards: SEMI E10, ISO/IEC 17025, JEDEC Standards
Semiconductor and Precision Electronics Reference

- **SEMI E10** – *Definition and Measurement of Equipment Reliability, Availability, and Maintainability (RAM)*
- **JEDEC JESD50** – *Guidelines for Calibration and Drift Compensation in Environmental Test Chambers*
- **ISO/IEC Guide 98-3 (GUM)** – *Guide to the Expression of Uncertainty in Measurement*

General Calibration and Metrology Standards

- **ISO/IEC 17025:2017** – *General requirements for the competence of testing and calibration laboratories*
  International standard for quality and technical requirements in laboratories.
- **ISO 10012:2003** – *Measurement management systems – Requirements for measurement processes and measuring equipment*
  Provides a framework for managing metrological confirmation and reducing drift.
- **NIST Handbook 150** – *NVLAP Procedures and General Requirements*
  U.S. guide to calibration traceability and laboratory accreditation.
- **ANSI/NCSL Z540.3-2006** – *Requirements for the Calibration of Measuring and Test Equipment*
  U.S.-based calibration system requirements tied to metrological traceability.

Interactive Workbook Section

Calibration Drift Blind Spot Workbook

This interactive workbook is designed to help identify hidden vulnerabilities in your calibration program. Each section presents a guided question with space to record findings, insights, and action plans.

1. Instrument Inventory Review

Prompt:
List all instruments currently in use and mark those not included in the formal calibration program.
Response Area:

____

____

____

____

2. Historical Drift Detection

Prompt:
Review past calibration records. Identify any instruments with >10% deviation over two or more intervals.
Response Area:

____

____

____

____

Michael D. Holloway. Accuracy and Calibration Failures in Science and Engineering, (291–294)

3. Environmental Influence Evaluation

Prompt:
List instruments exposed to extreme temperature, vibration, or moisture. Are they recalibrated more frequently?
Response Area:

4. Calibration Interval Assessment

Prompt:
Compare fixed vs. usage-based calibration intervals. Are high-use instruments recalibrated often enough?
Response Area:

5. Human Factors and Training

Prompt:
Survey technicians: Do they understand when to escalate a suspected drift issue? Are they trained to spot it?
Response Area:

6. SOP Compliance Check

Prompt:
Audit five recent calibration events. Were SOPs followed exactly? Were deviations documented and investigated?
Response Area:

____
____
____
____

7. Risk-Based Prioritization

Prompt:
Score instruments based on criticality and past drift. Are the highest-risk tools prioritized for short intervals?
Response Area:

____
____
____
____

8. Escalation and Communication Flow

Prompt:
Trace how a suspected drift case is reported and resolved. Is the process timely, documented, and reviewed?
Response Area:

____
____
____
____

Calibration Integrity Maturity Model & Scoring System

This tool complements the Calibration Drift Blind Spot Workbook. Use it to assess the maturity of your calibration program, score performance across key categories, and identify areas needing improvement.

Scoring Scale (Per Section)

Score	Maturity level	Description
3	Proactive excellence	Fully documented, consistently applied, reviewed, and continuously improved.
2	Compliant but reactive	Exists and functions, but reactive; some gaps in consistency or awareness.
1	*Ad hoc* or incomplete	Informal or inconsistently applied; depends on individuals.
0	Absent or unknown	No process, documentation, or known accountability.

Maturity Matrix Template

Use this matrix to score each section of your calibration program.

Section	Score (0–3)	Notes/improvement ideas
1. Instrument inventory review		
2. Historical drift detection		

(*Continued*)

Michael D. Holloway. Accuracy and Calibration Failures in Science and Engineering, (295–298)

(Continued)

Section	Score (0–3)	Notes/improvement ideas
3. Environmental influence evaluation		
4. Calibration interval assessment		
5. Human factors and training		
6. SOP Compliance Check		
7. Risk-based prioritization		
8. Escalation and communication flow		

Interpretation of the Total Score (Out of 24)

Score range	Maturity level	Action guidance
20–24	Best practice	Maintain current systems; consider benchmarking and mentoring other sites.
14–19	Competent but reactive	Focus on continuous improvement, training consistency, and predictive tools.
8–13	At risk	Identify gaps in documentation, ownership, and automation. Prioritize fixes.
0–7	High risk/ non-compliant	Urgently establish or rebuild foundational calibration practices and ownership.

Use Recommendations

- Use this model quarterly or annually during internal audits or QA reviews.
- Compare scores across departments or facilities to track calibration maturity over time.
- Use section scores to prioritize training, update SOPs, or justify investments in metrology tools.

Recalibration Tracking Worksheets

These worksheets are designed to track the calibration status, history, and drift patterns of critical instruments. They assist in trend analysis, scheduling, and risk mitigation for recalibration needs.

Instrument Calibration Log

Instrument ID	Location/ area	Last calibrated	Next due	Calibrated by	Remarks

Michael D. Holloway. Accuracy and Calibration Failures in Science and Engineering, (299–300)

Drift Trend Analysis Worksheet

Instrument ID	Cal. date	Measured value	Nominal value	Deviation (%)

Recalibration Action Tracker

Instrument ID	Drift noted	Action taken	Recal date	Approved by

Error Reporting & CAPA Analysis Templates

This document includes structured templates to standardize error reporting, investigation, and corrective and preventive action (CAPA) analysis within calibration and quality assurance workflows.

Error Reporting Template

Report ID	
Date/time of discovery	
Reported by	
Department/area	
Instrument ID/process	
Description of error	
Initial impact assessment	
Immediate actions taken	
Related calibration records	
Attachments or evidence	

Michael D. Holloway. Accuracy and Calibration Failures in Science and Engineering, (301–302)

Corrective & Preventive Action (CAPA) Template

CAPA ID	
Linked error report ID	
Root cause analysis (5 whys/fishbone)	
Corrective action (what was done to fix it)	
Preventive action (steps to prevent recurrence)	
Responsible person(s)	
Completion dates (corrective & preventive)	
Effectiveness verification plan	
Follow-up review date	
Approval/QA sign-off	

References and Source Documentation

Standards and Technical References

ISO/IEC 17025:2017 – General requirements for the competence of testing and calibration laboratories, International Organization for Standardization (ISO)

ISO 11171:2016 – Calibration of automatic particle counters for fluids, International Organization for Standardization (ISO)

ISO 4406:1999 – Hydraulic fluid power – Fluids – Method for coding the level of contamination, International Organization for Standardization (ISO)

ISO 9001:2015 – Quality management systems – Requirements, International Organization for Standardization (ISO)

ISO 31000:2018 – Risk management – Guidelines, International Organization for Standardization (ISO)

NIST SRM 2806a – Particles in Mineral Oil Suspension (20 mg/kg), U.S. National Institute of Standards and Technology (NIST)

ANSI/HFES 100 – Human Factors Engineering of Computer Workstations, Human Factors and Ergonomics Society (HFES)

IEC 61508 – Functional safety of electrical/electronic/programmable electronic safety-related systems, International Electrotechnical Commission (IEC)

Government and Investigative Reports

Rogers Commission Report (1986) – Report of the Presidential Commission on the Space Shuttle Challenger Accident, U.S. Government Printing Office

NRC Special Inquiry Report (1979) – Three Mile Island: A Report to the Commissioners and to the Public, U.S. Nuclear Regulatory Commission

NTSB Aviation Accident Report AAR-04/01 – Controlled Flight into Terrain, Korean Air Flight 801, U.S. National Transportation Safety Board

Michael D. Holloway. Accuracy and Calibration Failures in Science and Engineering, (303–308)

Deepwater Horizon Commission Report (2011) – National Commission on the BP Deepwater Horizon Oil Spill and Offshore Drilling
Fukushima Accident Summary Report (2015) – The Fukushima Daiichi Accident International Atomic Energy Agency (IAEA)
GAO-21-200 – COVID-19: Opportunities to Improve Federal Response and Pandemic Preparedness, U.S. Government Accountability Office
OECD Report (2010) – Understanding Financial Market Complexity, Organisation for Economic Co-operation and Development (OECD)

Scientific and Technical Articles

McLean, D. & Kruger, K. (2012). Revisiting the SRM 2806 Particle Count Calibration Transition: Lessons for the Future. STLE Annual Meeting Proceedings
Nelson, R. & Freedman, D. (2001). The Drift in Thermocouple Measurements in Gas Turbines. Journal of Aerospace Engineering, 14(3), 145–153.
Gollner, M. J., *et al.* (2018). Data Integrity and the Need for Model Skepticism in Environmental Monitoring. Environmental Science & Technology, 52(9), 4555–4564.
Leveson, N. (2004). A New Accident Model for Engineering Safer Systems. Safety Science, 42(4), 237–270.
Kahneman, D., Lovallo, D., & Sibony, O. (2011). Before You Make That Big Decision... Harvard Business Review, June 2011.

Manuscripts Referenced

Leveson, Nancy. Engineering a Safer World: Systems Thinking Applied to Safety. MIT Press, 2012.
Dekker, Sidney. The Field Guide to Understanding 'Human Error'. CRC Press, 2014.
Taleb, Nassim Nicholas. The Black Swan: The Impact of the Highly Improbable. Random House, 2007.
Reason, James. Managing the Risks of Organizational Accidents. Ashgate Publishing, 1997.
Petroski, Henry. To Engineer is Human: The Role of Failure in Successful Design. Vintage, 1992.

Noteworthy Tools and Frameworks Cited

ALCOA+ Data Integrity Framework – U.S. FDA and WHO
FMEA (Failure Mode and Effects Analysis) – SAE J1739
HAZOP (Hazard and Operability Study) – IEC 61882
Lessons Learned Registers – Best practices from NASA, DOE, and IAEA
Monte Carlo Simulation – Quantitative Risk Modeling Techniques
Red Teaming Methodology – U.S. Department of Defense and NIST Frameworks
Bayesian Model Validation – Use in epidemiology, finance, and environmental systems

Case Study References

Thermocouple Drift in Aerospace and Power Industries

Thermocouples are widely used in high-temperature environments such as aerospace and power generation. Over time, they can experience drift due to factors like oxidation, metallurgical changes, and contamination. These changes can lead to inaccuracies in temperature measurements.ccpi-europe.com
Reference: What causes drift in Thermocouples? | News - CCPI Europeccpi-europe.com
https://ccpi-europe.com/2018/01/11/causes-drift-thermocouples/

Pressure Transducer Zero Drift (BP Oil Spill Related)

During the Deepwater Horizon oil spill, the accuracy of pressure transducers was critical for monitoring and controlling the subsea environment. Zero drift in these sensors could lead to significant errors in pressure readings, affecting decision-making during the crisis. Naval Sea Systems Command+1PNAS+1
Reference: Deepwater Horizon Oil Spill ResponseNaval Sea Systems Command
SH285-AA-RPT-010
Chrome-extension://efaidnbmnnnibpcajpcglclefindmkaj/https://www.navsea.navy.mil/Portals/103/Documents/SUPSALV/SalvageReports/Deepwater%20Horizon.pdf

The pH Electrode Calibration (Environmental Monitoring)

pH electrodes are essential for monitoring environmental water quality. Calibration drift can occur due to factors like electrode aging, contamination, and temperature changes, leading to inaccurate pH readings.
Reference: EPA Method 150.1: Measurement of pH in Water

CT Scanner Dosimetry Drift (Medical Imaging)

CT scanners rely on precise dosimetry to ensure patient safety. Drift in dosimetry calibration can result from sensor aging or software updates, potentially leading to incorrect radiation dose delivery.
Reference: FDA: Reducing Radiation from Medical X-rays

Mass Standards Drift (Pre-Kilogram Redefinition)

The International Prototype of the Kilogram (IPK) was found to drift over time due to surface contamination and material loss, prompting a redefinition of the kilogram based on fundamental constants.
Reference: BIPM: The Kilogram Redefinition

Gas Chromatograph (GC) Flame Ionization Detector (FID) Calibration Drift

FID detectors in GC systems can experience calibration drift due to factors like contamination, aging components, and temperature fluctuations, affecting the accuracy of hydrocarbon measurements.
Reference: EPA Method 8015: Nonhalogenated Organics Using GC/FID

Optical Emission Spectrometer (OES) Metal Analysis Drift

OES instruments can drift over time due to factors like electrode wear, optical component degradation, and environmental conditions, impacting the accuracy of metal composition analysis.
Reference: ASTM E1086-14: Standard Test Method for OES of Stainless Steel

Industrial Scales – Load Cell Drift

Load cells in industrial scales can drift due to temperature changes, mechanical stress, and aging, leading to inaccurate weight measurements.

Reference: NIST Handbook 44: Specifications and Tolerances for Weighing Devices

Spectrophotometer Baseline Shift in Color Measurement

Spectrophotometers can experience baseline shifts due to lamp aging, detector drift, and environmental factors, affecting color measurement accuracy.
Reference: ASTM E1164-12: Standard Practice for Obtaining Spectrometric Data

Flow Meter Calibration Drift in Custody Transfer

Flow meters used in custody transfer can drift over time due to wear, fouling, and environmental conditions, leading to measurement inaccuracies in fluid transfer.
Reference: API MPMS Chapter 5: Measurement of Liquid Hydrocarbons by Flowmeters

Ion Selective Electrode (ISE) Calibration for Sodium and Chloride in Water

ISEs can drift due to membrane degradation, contamination, and temperature changes, affecting the accuracy of ion concentration measurements in water.
Reference: EPA Method 9211: Potentiometric Determination of Chloride

Infrared Thermometer Drift (Non-Contact Thermometers)

Non-contact infrared thermometers can drift due to sensor aging, environmental factors, and calibration issues, leading to inaccurate temperature readings.
Reference: FDA: Infrared Thermometers for Human Temperature Measurement

Index

Also of Interest

By the same author

The Analytical Chemistry Laboratory Companion, by Michael D. Holloway, ISBN:9781394355082. *The Analytical Chemistry Laboratory Companion* is essential for both students and professionals, as it provides quick, clear explanations on critical topics in analytical chemistry, equipping you with the statistical tools necessary to ensure accurate and reliable data interpretation.

Dictionary of Industrial Terminology 2nd Edition, by Michael D. Holloway and Emma Holloway, ISBN: 9781119363446. Greatly expanded and revised, this second edition is the most comprehensive and in-depth dictionary of industrial terms available, a must-have for any engineer, scientist, technician, or operator working in industry.

Fracking 2nd Edition: Further Investigations into the Environmental Considerations and Operations of Hydraulic Fracturing, by Michael D. Holloway, ISBN: 9781119363422. This revised, expanded edition of an instant classic explores the history, techniques, and materials used in the practice of induced hydraulic fracturing, while examining the environmental and economic impact.

Check out these other related titles from Scrivener Publishing

Environmental Analysis Laboratory Handbook, edited by Anshul Nigam and Rupal Gupta, ISBN: 9781119724803. The most comprehensive and up-to-date volume on environmental analysis available today, this is the standard laboratory reference for any environmental or chemical engineer, chemist, or scientist.

Integrating Green Chemistry and Sustainable Engineering, edited by Shahid-ul-Islam, ISBN9781119509820. This groundbreaking book covers the recent advances in sustainable technologies and developments, and describes how green chemistry and engineering practices are being applied and integrated in various industrial sectors.

Chemical Process Engineering Volume 1: Design, Analysis, Simulation, Integration, and Problem Solving with Microsoft Excel-UniSim Software for Chemical Engineers Computation, Physical Property, Fluid Flow, Equipment & Instrument Sizing, Pumps & Compressors, Mass Transfer, by A. Kayode Coker and Rahmat Sotudeh-Gharebagh, ISBN 9781119510185. Written by one of the most prolific and respected chemical engineers in the world and his co-author, also a well-known and respected engineer, this two-volume set is the "new standard" in the industry, offering engineers and students alike the most up-do-date, comprehensive, and state-of-the-art coverage of processes and best practices in the field today.

Chemical Process Engineering Volume 2: Design, Analysis, Simulation, Integration, and Problem Solving with Microsoft Excel-UniSim Software for Chemical Engineers, Heat Transfer and Integration, Process Safety, Chemical Kinetics and Reactor Design, Engineering Economics, Optimization, by A. Kayode Coker and Rahmat Sotudeh-Gharebagh, ISBN: 9781119853992. Written by one of the most prolific and respected chemical engineers in the world and his co-author, also a well-known and respected engineer, this two-volume set is the "new standard" in the industry, offering engineers and students alike the most up-do-date, comprehensive, and state-of-the-art coverage of processes and best practices in the field today.

Petroleum Refining Design and Applications Handbook Volume 1, by A. Kayode Coker, ISBN: 9781118233696. The most comprehensive and up-to-date coverage of the advances of petroleum refining designs and applications, written by one of the world's most well-known process engineers, this is a must-have for any chemical, process, or petroleum engineer.

Petroleum Refining Design and Applications Handbook Volume 2: Rules of Thumb, Process Planning, Scheduling, and Flowsheet Design, Process Piping Design, Pumps, Compressors, and Process Safety Incidents, by A. Kayode Coker, ISBN: 9781119476412. The second of a three-volume set of the most comprehensive and up-to-date coverage of the advances of petroleum refining designs and applications, written by one of the world's most well-known process engineers, this is a must-have for any chemical, process, or petroleum engineer.

Petroleum Refining Design and Applications Handbook Volume 3: Mechanical Separation, Distillation, Packed Towers, Liquid-Liquid Extraction, Heat Transfer and Process Safety Incidents, by A. Kayode Coker, ISBN: 9781119794868. The third volume of a four-volume set of the most comprehensive and up-to-date coverage of the advances of petroleum refining designs and applications, written by one of the world's most well-known process engineers, this is a must-have for any chemical, process, or petroleum engineer.

Petroleum Refining Design and Applications Handbook Volume 4: Heat Transfer, Pinch Analysis and Process Safety Incidents, by A. Kayode Coker, ISBN: 9781119827528. The fourth volume of a multi-volume set of the most comprehensive and up-to-date coverage of the advances of petroleum refining designs and applications, written by one of the world's most well-known process engineers, this is a must-have for any chemical, process, or petroleum engineer.

Petroleum Refining Design and Applications Handbook Volume 5: Pressure Relieving Devices, Process Safety and Energy Management, Petroleum Product Blending, Economic Evaluation, and Sustainability, by A. Kayode Coker, ISBN: 9781394206988. The fifth volume of a multi-volume set of the most comprehensive and up-to-date coverage of the advances of petroleum refining designs and applications, written by one of the world's most well-known process engineers, this is a must-have for any chemical, process, or petroleum engineer.

Printed and bound by CPI Group (UK) Ltd, Croydon, CR0 4YY
17/09/2026
14964285-0002